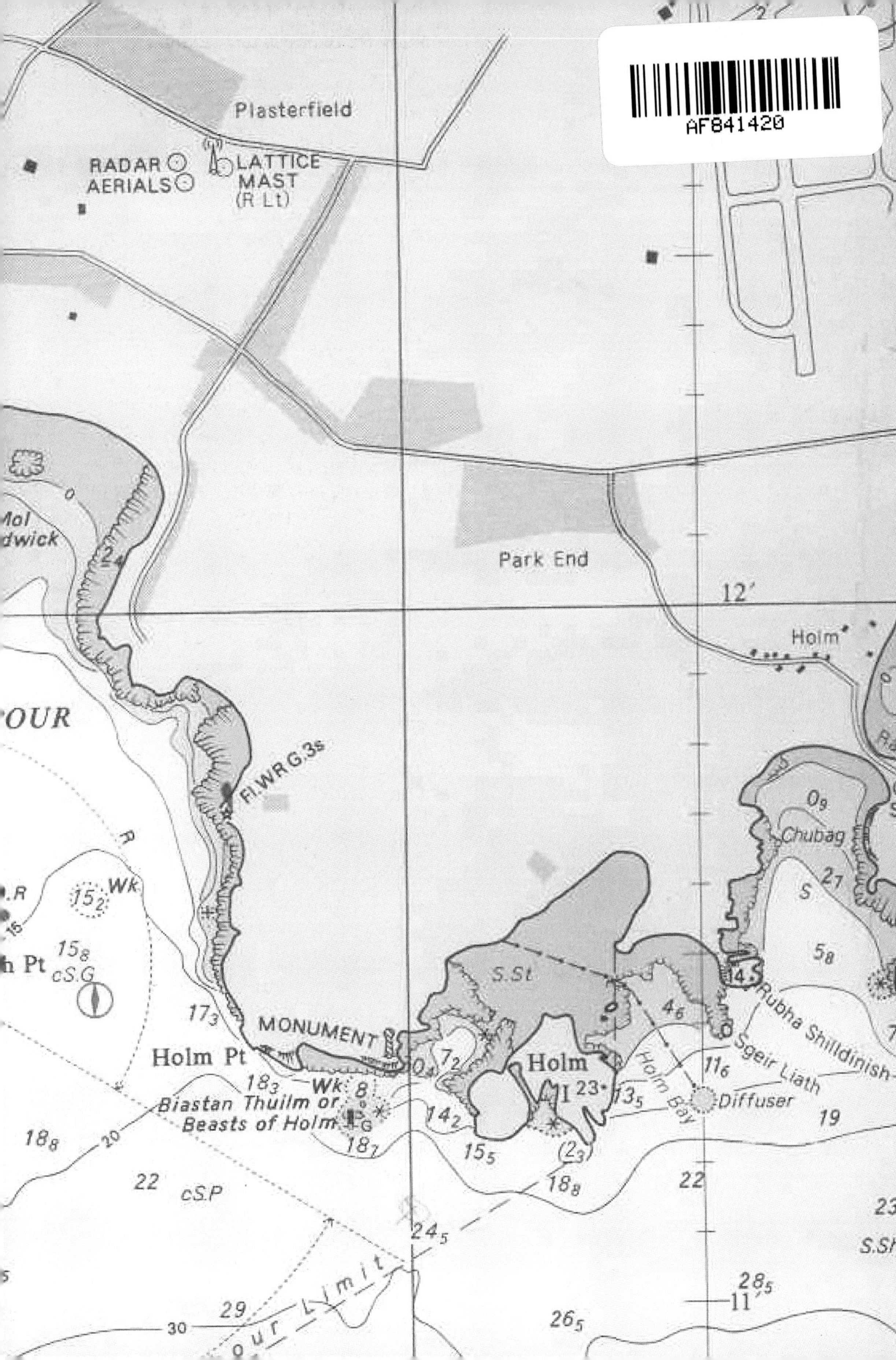
Plasterfield
RADAR AERIALS
LATTICE MAST (R Lt)
Park End
12′
Holm
Mol
dwick
OUR
Fl.WRG.3s
Wk
h Pt
cS.G
MONUMENT
Holm Pt
Wk
Biastan Thuilm or Beasts of Holm
S.St
Holm I
Holm Bay
Diffuser
Sgeir Liath
Rubha Shilldinish
Chubag
cS.P
S.Sh
11′
our Limit

It must be Stornoway

The Story of Stornoway Pier and Harbour Commission 1865 to 2004

Catherine Mackay

ARGYLL PUBLISHING

Argyll Publishing
Glendaruel
Argyll PA22 3AE
www.argyllpublishing.com

**British Library Cataloguing-in-Publication Data.
A catalogue record for this book is available from the British Library.**

ISBN 978 1 906134 1 1

Printing: Bell & Bain Ltd, Glasgow

Contents

Stornoway Gazette

Changing times in Stornoway harbour are reflected in this view of a modern Caledonian MacBrayne ferry, *Isle of Mull* and the traditionally-built sgoth Niseach *An Sulaire*

Foreword

Most harbours, and particularly Island harbours, are an exit and entrance to and from the outside world through which innumerable souls have passed over the decades, be they seamen, fishermen or the travelling public. Stornoway harbour is such a facility and is well known to all Islanders and to most visitors. Few, however, give much thought as to how, or indeed why, the port was established and has since been regulated, improved, and managed.

Stornoway Pier & Harbour Commission, now the Stornoway Port Authority, is the oldest extant statutory body in the Western Isles dating back to its formation in 1865, its original remit being to set up, commission and regulate harbour facilities. In May 2004 Stornoway Pier & Harbour Commission was reconstituted in terms of the Government's 'Modernising Trust Ports' initiative and, by means of the Stornoway Harbour Revision (Constitution) Order 2003, the Stornoway Port Authority was established.

It was felt that this occasion should be marked by recording the history of the Stornoway Pier & Harbour Commission (1865 to 2004) in book form so that as many as are interested may have the opportunity to read about the inception of the body and enjoy a flavour of its life and times. To this end a small but enthusiastic Editorial Committee was formed and, ably led by the Harbour Collector, Catherine Mackay, work began in gathering, collating and editing a vast amount of material.

I am honoured to have the opportunity to introduce this book which records the story of 139 years of the Harbour Commission's continuous service to our Island community, each generation of which has experienced enormous change in methods of transporting goods and people as well as ever increasing regulation pertaining to the Ports industry. Every year, many thousands of passengers and thousands of tons of cargo are safely and efficiently moved through the port, bearing testament to the dedicated professionalism of our harbour staff, to whom I pay tribute for their commitment and loyalty.

Over the years the Port has sought to stimulate local business and to support enterprise. Strategically it provides a good barometer of the economic wellbeing of the wider Hebridean community. Perhaps the greatest influences on its own activity and prosperity have come from without, eg emigration, two World Wars and the changing fortunes of various industries. Throughout all of these changes Stornoway harbour has risen to the many differing challenges and has evolved to its present standing as the principal port and gateway to the Western Isles.

My early memories of Stornoway harbour are of going there on evenings to await the arrival of MV *Loch Seaforth*. It was a place of cargo containers, wool bales and milk churns and with the fish mart close by, had an atmosphere of mystery and excitement. On the pier could be found a representative cross section of island dwellers; those who had come to meet relatives, and those who had come to collect goods, milk, newspapers and perishable foodstuffs mingled with casual onlookers, dockers, shipping company staff and port operatives. In my own case we waited for livestock to be walked ashore down gang planks; then we set off, herding them through the streets of Stornoway, to Ropework Park and often thence to Steinish, hail, rain or shine, summer and winter! I could never have imagined that some fifty years later I would have the responsibility and privilege of being Chairman of the Stornoway Port Authority.

The Authority is fortunate in having within its membership a group of talented and enthusiastic individuals who have given willingly of many hours of their time in the production of this book. A great deal of thanks are due in particular to Miss Mackay, past Chairman Sandy Matheson and past Vice Chairman Malcolm Macfarlane. Thanks also to Port Chief Executive John J Maclennan and his fellow Board members, both past and present, for their support of the project. The Authority is also grateful to Mrs Annice Macleod, Mrs Mary Mackenzie, and Mr Norman Macgregor for giving of their time and expertise.

Drawing on materials researched from Harbour Commission Minute Books and Records, newspapers from within the Highlands and Islands, various contemporary books and publications and with valued input from local contributors *It Must Be Stornoway*, in word and picture, provides a fascinating insight into life in and around the harbour and town, charts the development of the port infrastructure and airs tales of the politics of a local institution.

In commending this publication to you I hope that you will find it readable and experience as much pleasure in your reading of it as the 'Book Team' so evidently did in their composition of it.

Iain Macleod
Chairman, Stornoway Port Authority

Acknowledgements

The Port Authority is indebted to a host of individuals and organisations for assistance in the production of this history of Stornoway Pier and Harbour Commission. In particular the Authority is grateful to Historic Scotland for a grant from its Sponsorship Fund and to the Stornoway Historical Society for its donation.

Thanks for advice, photographs and information are due to Peter Bachelor; Iain Campbell; Donald Chapman; Peter Cunningham; Calum Ferguson; Florence Finlayson; Kinloch Historical Society; Lodge Fortrose; Rodney Long; Angus McCormack; Dierdre Macdonald; John Macdonald; Kate Macdonald; Malcolm Macdonald; Norman Macgregor; Chirsty Maciver; Sheriff C Scott and Mrs Christeen Mackenzie; Helen Mackenzie; Mary Mackenzie; Norman Mackenzie; Robin Mackenzie; Sandra Mackenzie; John Maclean Photography; Michael Maclennan; Mairi Maclennan; Annice Macleod; Charles Macleod; Murdoch Macleod: John Macqueen; Professor Donald Meek; Tony Morrison; Frances Murray; Museum nan Eilean; Charles Nicolson; Tony Robson; Scottish National Archives; Angus Smith Photography; James Smith Photography; Margaret Smith; Morag Smith; Stornoway Historical Society; Anne Sutherland; William Watson; Western Isles Libraries; Western Isles Registration Service; the families of Julia Fraser and Alasdair Macleod; and the archives of *Highland News*, *The Scotsman* and *Stornoway Gazette* newspapers.

To the many others in the community who have offered support, encouragement and help in the compilation of this book, Stornoway Port Authority expresses sincere thanks.

Stornoway Harbour – a print of a watercolour by William Daniell, 1819

'Soon there was no doubt about it being Stornoway. Beyond a green island in the wide fairway, they saw the piers about a broad rectangular promontory, crowded wih buildings that stuck far out into the water, the shore curving away to either side from its root. At anchor off this promontory, Finn counted nine vessels – smacks, schooners, and no less than two three-masted barks. Straight over them, on the treeless slopes across the bay, stood a round tower, and over it to the right, Stornoway Lodge, the residence of the island's propietor.'

Neil Gunn **The Silver Darlings**

Introduction

The name Stornoway is derived from two Icelandic (Old Norse) words: *stjorna* denoting steering or anchor and *vagr* signifying a bay or a creek.

It may readily be concluded from that name alone, 'anchor bay', that some form of maritime settlement has existed here since shortly after the Norse invasion of the Hebrides in 878AD. Thus the genesis of Stornoway harbour is rooted in the Viking expansion along the seaboard of North West Europe. The southern and western thrust of this Norse colonisation also crossed the ocean to Shetland, Orkney and the Hebrides and as far south as the Isle of Man and Dublin.

The bay at Stornoway provided all the essential elements for land-based winter quarters required by the Viking settlers. Sheltered by the embracing arms of Arnish and Holm, with deep water and accessible at all states of the tide, the bay was like any Norse fjord. A flat sand bar at what is now the Point Street peninsula not only protected an inner harbour but also gave ideal conditions to beach and careen the longships for repair. The water-courses at the Creed and Bayhead gave ample supply of fresh water and flat land at Inaclete and Bayhead provided for cultivation and camp building to provision and shelter colonists and settlers. Thus, more than one thousand years ago, Stornoway evolved as an important base and staging post in the sea-borne Viking empire.

Evidence of a fortified settlement at the haven in Stornoway emerged by the end of the twelfth century when a Norse warrior chief, Nicol Hasson, built a fortified citadel on an islet in the bay. This fort, later called Stornoway Castle and reduced to ruins by Cromwellian troops in the seventeenth century, was commemorated on a plaque on the north east corner of the Maritime Buildings. The derelict remains of the fort were covered over in 1886 when the first section of Number 1 Pier was erected.

The Norse supremacy ended in 1265 and by the end of the thirteenth century, Stornoway was, at least nominally, fully integrated into the Kingdom of Scotland. For the next three centuries, Stornoway and its harbour were of little consequence but towards the end of the sixteenth century a report to King James VI stated, 'The Isle of Lewis is very profitable and fertile in allsorts of corn, beasts, wildfowl and fishes.' This recognition of the superabundance of fish led to a strong trading connection with Holland and in 1594 Dutch fishermen made their first recorded visit to Stornoway.

Four years later the 'Fife Adventurers' made the first of two abortive efforts to colonise Stornoway for the Crown because the predominant clan, the Macleods of Lewis, refused to submit to King James VI. However internecine squabbling and bloodshed within the Macleod clan allowed the Mackenzies of Kintail to reach agreement with King James and from that time in 1611, when the Mackenzies were granted a Charter of Burgh and Barony for Stornoway, the future of the town and its harbour was secured.

During the next two and a half centuries Stornoway harbour was of both military and commercial importance in the affairs of Scotland and Britain. After the English Civil War and during the time of the Protectorate, Oliver Cromwell feared that the Dutch, in alliance with the exiled King Charles II, would stage an invasion through the back door of Britain, the north west of Scotland and, accordingly, he set up a military base in Stornoway harbour. During this Cromwellian period the old castle in Stornoway Bay was demolished and a new fort built on the peninsula bounded by what is now Cromwell Street, North Beach and South Beach. From the time of the Restoration of the monarchy in 1660 and throughout the next one hundred and eighty years of Mackenzie ownership of Lewis the development of the fishing industry became the main impetus for the gradual development and improvement of Stornoway harbour.

At the opening of the eighteenth century, in 1720, the merchants of Stornoway had Zachary Macaulay, the Chamberlain of Lewis, act as agent for Bailie Stewart of Inverness; at that time all the exports were carried in vessels hired from the mainland but by 1760 this had changed and all exports were shipped in local vessels with all the added value for Stornoway that this implied. The census of 1763 showed that 500 men were employed in fishing from Stornoway, where there were 50 fishing boats and by 1793, 10,945 barrels of herring and 68 tons of white fish were exported from the harbour.

In the early part of the nineteenth century the decline of the kelp industry at the end of the Napoleonic Wars created a significant increase in poverty and hardship in the island and, although the white fish industry maintained its value, the herring fishery out of Stornoway stagnated because neither the nets nor the boats were suitable for deep sea fishing in the Minch. It was not until toward mid century that the introduction of a larger boat, known as the 'Anstrutherach', caused a revival in the herring fishery out of Stornoway.

This coincided with another no less significant event. In 1844 James Matheson a merchant prince of Highland birth, who had made a vast fortune in the China trade, purchased Lewis from the by now bankrupt estate of the Mackenzies of Seaforth. Although in earlier times the Matheson family had been keen to help the fishing industry, Sir James, as he became in 1851, was not personally interested in the fishing business, believing that the capital for investment should come from the fishing entrepreneurs rather than himself.

Sir James recognised and acted upon the need to improve the facilities at Stornoway harbour on which nothing had been expended since 1818. Although he was more interested in land improvements, Sir James did provide the capital for and had put in hand a number of projects. These gave substantial local employment and were also a source of more indirect employment, municipal progress and social and educational advancement. Those directly related to the harbour included: the purchase and revenue subsidy for the steam packet *Mary Jane*, running between Stornoway and Ullapool; the construction of new and the repair and extension of old piers; the provision of the Patent Slip and Ship Building Yard at Iomairsligeach (Shell Street); the setting out of land for fish curing stations and, finally, in order to settle the vexed 'foreshore question' he had the constitution of The Stornoway Pier and Harbour Commission promulgated by Act of Parliament in 1865.

Sandy Matheson

Charles Nicolson

Lews Castle which overlooks Stornoway Harbour.
In 1844 the wealthy trader James Matheson, co-founder of the Jardine Matheson Company, bought the island from the bankrupt estate of the Mackenzies of Seaforth. Designed by architect Charles Wilson, the Castle was built on the site of Seaforth Lodge and took seven years to complete

The Beginning 1864–1875 1

ON AN AUGUST day in 1864 Sir James Matheson, in the presence of William Edward Martin, his butler and Thomas Mackay, his piper, signed the feu charter bringing Stornoway Pier and Harbour Commission into being. His view from Lews Castle windows down into Stornoway Harbour was very different from today.

A seafarer entering the harbour in 1864 would find a shore line of sand and gravel running along the south side of South Beach Street with no quays at all on that side of the harbour. Coming nearer he would see the ruins of the old Macleod castle in the area where the recently demolished Maritime Buildings once stood and slightly to the north of that, somewhere in the area of the present day car parks on No 1 Pier,

Sir James Matheson, a native of Sutherland, became a hugely successful trader in opium and tea and was credited with the building of Hong Kong harbour. His vast wealth enabled him to purchase the estates of Achany and the Lews in the middle of the 19th century. The Matheson ownership of Stornoway Harbour sometimes proved to be stormy.

would be the Steamer Quay built by Sir James for the thrice weekly steamer connection to the mainland. Further along in the area which is now King Edward's Wharf , the old barque *Amity* did duty as another pier.

A seawall ran along a gravel beach on the western side of Esplanade Road close to the present Port Authority offices in Amity House, until it reached the oddly shaped 'Big Quay' at the end of Esplanade Road. This quay extended about 100 feet westward from the junction of North Beach and Esplanade Road and eastward to the end of Quay Street roughly in the area in front of the Old Sail Loft.

Turning into the Inner Harbour the sailor would find the shore line running along North Beach and Cromwell Street. If it was Spring tides, there might be vessels beached where Perceval Square is today, with their bowsprits jutting out into the main thoroughfare on Cromwell Street. The Bayhead river estuary was open right up to the Porter's Lodge at the entrance to the Castle policies. Along the shore line there were several little piers built and owned by private individuals, carrying on their own private trade. Fishermen might be engaged in 'barking' their nets on the foreshore, a messy business entailing the boiling of a substance called cutch made from the bark of the acacia tree, imported from the East. The ropemakers might be boiling tar in which to soak their ropes and the odours from these might be making their presence felt up at the Castle if the wind wafted them in that direction.

It was mainly the sights, sounds and smells of the growing fish trade so close to the Castle policies that had prompted Sir James to approach the Treasury and offer to purchase the whole of the foreshore of Stornoway Harbour. His petition to the Government stated that his main object was the promotion of the public convenience and advantage. He complained that feuars were able:

> 'to occupy and build upon the foreshores, each man according as it suits his fancy or as he may deem it conducive to his own interest, and to convert the foreshore thus unwarrantably occupied, to purposes which seriously interfere with the public thoroughfare and convenience.'

He had no wish, Sir James said, to secure any personal benefit whatsoever and he felt that, as proprietor of the island and of the burgh, he was. . .

> 'in a situation which fairly entitles him to expect a grant of the foreshores of the bay of Stornoway, from Arnish Lighthouse on the west to Upper Holm on the east.'

Sir James further submitted to the Lords Commissioners that, since the foreshores in question were incapable of being turned to any directly profitable or rent-yielding purposes, therefore the price ought to be an extremely moderate one.

Their Lordships handed the matter over to the Commissioners for Woods and Forests (now the Crown Estate) whose legal adviser Mr Horne WS, Edinburgh asked Sir James to make an offer. Mr Horne felt that there was no need for a special survey and report since there was no pecuniary advantage to Sir James. He only wished to protect the lands against the encroachments of the sea or the acts of individuals. Mr Horne therefore recommended acceptance of his offer of £400.

The Foreshore Question

Thus the foreshores of Stornoway Harbour passed into the complete control of the Matheson estate. Sir James Matheson now had what he had asked for in his submission – the power to:

> 'grant titles on such conditions as shall tend to the good of the general community and enable him to prevent erections that not only interfere with that but which are turned to uses at once offensive in their own nature and calculated to impair the healthiness of the town.'

However, there were a number of individuals who saw things differently. They were the feuars and businessmen of Stornoway who had already been trying to obtain Parliamentary powers to manage the harbour properly. Some of them were members of the Quay Committee, established by public subscription almost fifty years earlier. They were as keen to have the harbour regulated as was Sir James but they saw it in the light of the business and fishing interests of the town. They believed that the growing fishing and shipping trade of the port was best served by a publicly managed harbour authority. By this time they had built up some funds and had been persuaded to approach Sir James for his help in Parliament in promoting a Private Bill.

They were taken unawares, therefore, by his acquisition of the foreshores and their response to this action began the tussle which became known as the 'Foreshore Question'. Those feuars who owned piers around the town opened their post one morning in January 1863 to find letters from Donald Munro, the Estate Factor, telling them they could no longer carry out any fish curing or boat building thereon, nor could they let them to others during the fishing season.

> 'I think it proper to apprise you that Sir James Matheson cannot in future recognise your right to let your pier for curing or boat-

> building purposes, or to execute any right or ownership over the same...'

Since they had borne the cost of erecting these piers themselves and had carried out their daily business on them for years, their blood was up and battle was joined.

A Public Meeting was called for 11th February 1863 in the Masonic Hall. It was chaired by Kenneth Smith, a forty-year old Stornoway merchant, known locally as 'California' due to his having spent a number of years in the USA. The meeting appointed a Provisional Committee of eight members to meet with Sir James the following day. They approached the matter with some deference, saying that they wished to fall in with his ideas for the good of the town and welcoming any additions to quay space and sanitary improvements. However they suggested that any improvements should be publicly managed.

The meeting with Sir James was not a success and the Committee, with the backing of the townspeople and now swelled by another eleven members, met the proprietor head on. They refused to accept that he had sole authority over the Public Quay, they would not hand over any accumulated funds and they offered to relieve him of the responsibility of providing more harbour accommodation if he would support their bid for an Act of Parliament.

Eventually Sir James raised a court action against 63 feuars of the burgh in an attempt to force them into handing over the Quay Committee funds and any dues collected and to accepting that he had sole right over the pier. The Committee engaged a legal adviser but knew their chances of success were slim. They went back to the public and asked for their backing to take the matter to the House of Commons.

Kenneth Smith stated that the feuars intended to appear before the Court as defenders in order to protect their interests. 'Are we tamely to allow ourselves to be deprived of our public rights? Those who went before us kept their ground nobly. Posterity would brand us with infamy if we were to surrender ours,' he ended to cheers.

Captain Donald Mackenzie, a local shipowner, moved as the first resolution that the grant of the foreshores did not convey to Sir James the powers and rights he claimed under it and that an Act of Parliament should be procured for improving, maintaining and managing the harbour by a competent and legally constituted harbour authority which would represent the general interest of the port. *The Scotsman* quotes him as saying:

> 'This is the first time in my recollection that the people of this town

> have been opposed to their proprietor. But I think they have the best of reasons for being so opposed now that an attempt is being made to wrest their prosperity from them. . .'

Commenting on Sir James' assertion that he had acquired the foreshore with a view to public benefit, Malcolm Mackenzie, a merchant, said that the public would not stoop so low as to benefit from people being deprived of their property built for their own and the public convenience and at their own expense. 'We are not a parcel of obstructives,' he said, 'but we would be no men if we did not defend public and private rights, in regard to improvements, which had been already made by the industry of our forefathers.'

Within a very few days a Petition was signed by 292 people calling for a Public Inquiry and as the 'Foreshore Question' became ever more public and unpopular, the pragmatic Sir James accepted a compromise in which he would retain some control of the harbour but in which the harbour users would also have their place.

Under way

Thus, after months of wrangling involving public meetings, legal exchanges and petitions to Parliament, Sir James and the feuars of Stornoway had now reached agreement on how the regulation and development of Stornoway Harbour was to be achieved. It was to be through the establishment of a Commission consisting of the three members elected by the townspeople, three appointed by the proprietor and one by the Sheriff. The proprietor was to be Chairman and in his absence his Chamberlain would take his place.

The Chamberlain in 1864 was Donald Munro, a name well known in Lewis history and the Minutes of the first recorded meeting in the autumn of 1864 show him in the Chair. At that meeting Peter Macfarlane was appointed Interim Harbour Master and Collector of Dues, and William Ross became Clerk and Treasurer *ad interim*.

Throughout the winter and following Spring, usually under the Chairmanship of Donald Munro, the Commissioners considered the terms of the Stornoway Harbour Order which would regulate the harbour affairs. They instructed the Edinburgh Agents, McCheyne and Stuart to immediately procure an Act of Parliament; they instructed the Clerk to 'obtain all possible information with as little delay as he can as to the cheapest mode of procuring plans and surveys of work similar to those in contemplation'; they employed Stevenson Engineers, Edinburgh, to look into the cost of the probable extension of the pier; and they agreed on a Schedule of Dues and Rates.

The Dues on Goods and Animals begins with Ale, Beer or Porter per Hogshead, through Eggs per dozen, Flour per sack, Guano per ton, Gunpowder per cwt, Tobacco per roll (one penny), Wheels (cart) per pair (twopence) and ends with Yarn with fifty or so other commodities in between. Vessels paid 3d per registered ton for loading or unloading.

On 29th June 1865 the Stornoway Harbour Order received Royal Assent and the Commissioners were in business. On that very day they began to record vessel movements in the harbour in the first of a series of leather bound ledgers which continued to be entered for over a century. The first ship recorded is the 45 ton *Star* landing barrels, registered in Fraserburgh and skippered by J Stephen. She paid 5s 7d harbour dues. Three other vessels are entered that day – the *Express* of Wick, also landing barrels, and two Jersey registered boats, the *Louisa Jane* and the *Dart*, one shipping herring, the other landing salt.

At this point there is no record of where ships came from or their destination, and the cargoes relate mostly to the fish trade, but the following year entries begin to show the arrival of cargoes of slate from Ballachulish, stones from the Isle of Martin, timber from Norway and all parts of Scotland. The *Favourite* of Arbroath brought coals from Newcastle in May 1867.

The Commissioners in 1865 were Kenneth Smith, 'California'; Daniel Macalister, Shipowner; Norman Maciver, Shipowner; Captain Donald Mackenzie, Shipowner; Kenneth Mackenzie, Bank Agent for the National Bank; and James Mackenzie, Merchant. The Harbour Order gave them powers to levy dues and rates, borrowing powers of up to £10,000 and defined the limits of Stornoway Harbour. Apart from putting in place a whole new structure for the day-to-day running of the harbour, they had before them the task of carrying out the new works authorised by the Order.

In the autumn they advertised for tenders from contractors to carry out the construction of a timber wharf of 150 feet in front of the Steamboat Wharf and a masonry quay wall running 350 feet northwards from that to join the Big Quay seaward of high water mark. The space between this quay and the high water mark was to be filled in to provide fish curing stances. The timber wharf became known locally as 'Quay an Amity' because it was built on that site; the masonry wall on Esplanade was known as the 'Stone Quay'

The successful tenderers were A & K Macdonald of Park House, Paisley Road, Glasgow with a price of £6734. The Clerk informed them that another offer had been received for considerably less money but that the Commissioners considered Macdonalds would proceed more promptly and energetically with the works. The contract is incorporated in the Minute Book with detailed instructions, from crop ties being of green heart and all the rest of the best Baltic timber to the iron work being of the best Govan iron.

Financing the works proved difficult since the Public Works Loan Board was unwilling to grant money to the new body. In 1867, therefore, Sir James Matheson provided the loan to make the work possible, allowing them to draw out the money as they required it to avoid paying overmuch interest. The final account of the contract work in 1869 showed that the whole cost of the new quays amounted to £8009 12s 2d.

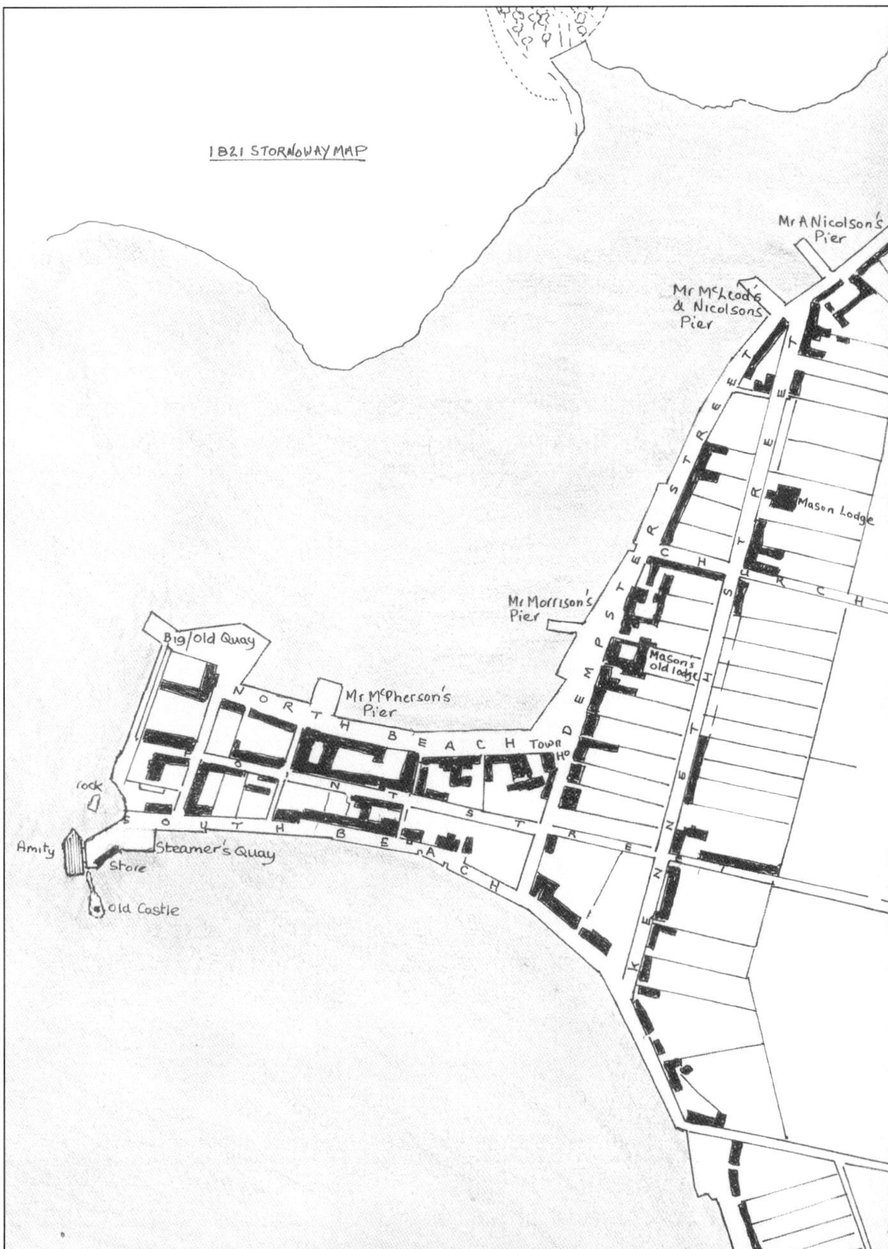

Stornoway Harbour as it was in the 1864 plan

Early Days

When the Harbour Order was confirmed, the Commissioners had advertised for a Harbour Master and Collector of Dues, Peter Macfarlane having only been temporarily appointed and, judging from entries in the Minute Books, obviously being relieved to relinquish his duties. William Lees, residing in Liverpool but of a Stornoway seafaring family, took up his post in August 1866 at a salary of £90 per annum and with a Bond of Caution running to three pages in the Minute Book. Alexander Morison, Junior, Merchant Fishcurer and Shipowner of Stornoway stood as his 'cautioner' in the sum of £200.

Mr Lees remained with the Commission for thirty years and for almost twenty of these he held the combined post of Harbour Master and Collector of Dues. In his resignation letter he says that 'for various reasons which it is perhaps needless to enumerate my duties during these years were very harassing' and indeed in the early days of the organisation, it must have been a difficult post to hold.

With the fishing industry in full swing and a harbour crammed with boats, having to apply new rates and rules to the skippers and curers around the pier would have been difficult enough. To have to do it under the watchful eyes of seven new Commissioners who had an agenda for improvements and who included the exacting Mr Munro, must have been especially taxing.

In April 1867 he was allowed to hire a clerk to help him during the fishing season at six shillings a week. He was also given the aid of two extra policemen for the season – Murdo Macleod and Malcolm Smith were sworn in as special constables in May – and he had the guidance of a formidable set of byelaws. If any Shipmaster or other person interrupted, hindered, threatened or insulted him in the execution of his duty they were liable to a fine not exceeding £5. Captain Begg of the schooner *Prince Alfred* tried it and found himself in front of the Sheriff Substitute the following day.

More help came his way in May 1867 when Mr John Groom, the Collector of Customs, offered to collect harbour dues from vessels at a commission fee of two and half per cent of the collections. The Commissioners accepted the offer. Both Mr Groom and Mr Lees found themselves having to explain to the Commissioners at the end of the fishing season why 128 out of the 552 herring fishing boats landing fish 'had been allowed to escape without paying'. Their answer is unrecorded but the meeting reminded them of the urgent necessity to collect dues as the boats arrived.

That summer Captain Macdonald of the *Princess Royal* wrote to the Commission regarding the placing of buoys in the harbour. He considered it desirable to place a beacon on the south edge of the Big Rock off Sandwick Bay, which in southerly to south westerly gales covered at half flood. He listed seven boats which to his knowledge had grounded on it in the last four years. He also mentioned the 'Seir Rock' (*sic*) a continuation of the north end of Arnish Rock and drying at low water of Spring tides which, he said, was not so dangerous but vessels were occasionally stranded on it. The schooner *Surprise* had recently been lost on it and what he described as the 'larger class of fishing boats' were frequently damaged by touching it. He continues prophetically–

> 'There are also very dangerous rocks known as the Beasts of Holm, about two cables length SE of Holm Point to which I beg to draw the attention of the Commissioners; they be very much in the way of vessels entering or leaving the harbour, and hugging the east shore, cover at half flood and several casualties have occurred on them.'

William Ross, Clerk to the Commission, was instructed to report the contents of Captain Macdonald's letter to Messrs Stevenson, their Edinburgh engineers. The Lighthouse Commissioners of Northern Lights later approached Sir James offering to build a beacon on one side of the Harbour if the Harbour Commission would build one on the other but the Commissioners considered their finances could not stretch to it at that stage.

Later that year Ross was instructed to have levels taken at the Bayhead River with a view to diverting it in a direct line to the Pier at the Old National Bank and thence by North Beach to the Big Quay. It was considered that this would help the sanitary condition of the town as well as improve the harbour.

A Committee of three Commissioners was appointed in 1870 to find means to obviate the problem of a number of idle boys rushing on board the steamer and blocking the passage to the annoyance of the travelling public. Their answer was to suggest a railing $5^1/_2$ feet high and sharp pointed at the tip to keep out all but those who had business at the steamer until the passengers landed.

Settling In

After the first flurry of activity in setting up the Commission and getting the new works started, the meetings settled to a less frequent pattern. This might be because the once all-powerful Munro had his hands full with the unrest in the crofting communities at the time, culminating in the Bernera Riot in 1874 and his resignation as Chamberlain in 1875, when his place was taken by William Mackay.

William Lees, the Harbour Master, continued to shoulder responsibility for the day-to-day operation of the port. The phrases 'the Harbour Master was instructed' or – more ominously – 'was reminded' appear frequently in the Minutes. During the busy seasons he had an assistant to help and a clerk, one of whom for a time was William Cook Mackenzie of later fame as a nationally known author and local historian.

The position of Harbour Master seems to have been a thankless one. Because at the time the bulk of the trade was done by sailing vessels, he had to attend the tides late and early in all weathers and the limited accommodation space meant that he had to deal with recalcitrant skippers refusing to move their vessels, frequently leading to the imposition of fines.

He had to find berths for the increasing steamer traffic and fishing trade while the new works went on. He was reminded of the necessity of making sure all berths were free from obstructions when a stone went through the hull of Mrs Isabella Robertson's boat and the Commission had to re-imburse her to the tune of over £10.

He had to bring to book those such as the Master of the *Lady of the Lake* who discharged ballast into the harbour and left £2 with the Clerk before he departed, which the Commissioners directed to be banked and the remainder of the fine to be retrieved should the vessel ever return. He had to do something to prevent unfortunate accidents such as that which occurred when Mr Falconer, Free Church Probationer, fell over the new pier. He had to make sure the new timber wharf was covered with planks to avoid the horses' hooves damaging it as they carted away around 500 tons of coal from the German barque *Dr Peterman*, alongside for repairs – and he had to collect harbour dues from the hundreds of fishing boats filling the harbour in the busy season.

With the new quay space had come increased traffic and new storage sheds were constructed to accommodate the cargoes being landed from steamers beginning to call regularly at the port. Many of the Commissioners' decisions had to do with the setting of charges on these regular callers, the most regular of which was the *Ondine*, the mail steamer introduced by Sir James Matheson which ran between Ullapool and Stornoway. Her first proper entry in the Arrivals Book is in 1872 although William Lees has a pencilled note in the back of the ledger saying she arrived with first mails on 4th August 1871.

The Commissioners offered Sir James free berthing for the vessel in consideration of 'the great boon it has been to the whole community to have such a regular mail communication'. He graciously declined the offer, saying the vessel should pay a charge in proportion to that paid by Messrs Hutcheson on their steamers which was £90 for the half year. The *Ondine's* was eventually set at £20 per annum.

Mr Douglas of the Highland Railway Company wrote to ask for quay accommodation for their vessel the *Ferret* which it was intended should ply between Strome Ferry and Stornoway. Dues were set at £80 which he protested about to no avail. Mr J M Morison, the local steamboat agent for Messrs Langlands, had two vessels, the *Tusker* and the *Fairy Queen* at a compounded sum of £80. David MacBrayne appears in the record as well, although at this stage in a terse note by the auditor at the end of the accounts –

> 'The sum due by Mr MacBrayne for his steamers and as rents are every year in arrears which he should be called upon to avoid in future.'

John Maclean

John Macqueen

Fishing Boats 1890s – the days when, according to J N Maciver, 'the bay was a leafless forest of masts at the weekend and a fluttering of rich brown sails spreading their wings seaward on a fair June morning'.

Expansion 1876–1892 2

TOWARDS THE END of the 1870s the Commissioners, by now including another notable name, Aeneas Mackenzie, were set on further extensions to the quay space, although they still owed a debt of over £7,000 to Sir James. The growth of the fishing industry and the use of larger vessels were putting pressure on the limited quay space available. In 1875 the number of cod and ling cured in the Stornoway district was 471,611 and the barrels of herring shipped from the port totalled 57,709.

The following year the Burgh Commissioners complained about the state of the Inner Harbour near the Post Office and Donald Morison, Creed, was directed to report on the probable cost of erecting a new wall from the end of William Morison's quay to the end of the Lews Hotel Quay and also the cost of diverting the Bayhead River. The cost was estimated at £740 and a Mr Caunter whose address is given as Millburn Cottage, Stornoway (now Glen House) was asked to draw up plans to submit to the Board of Trade. The Harbour Commission began to look into the possibility of extending the timber wharf round the old Castle as originally discussed and decided to go ahead with Mr Caunter's stone wall along Cromwell Street Quay.

Finding the money

The Commissioners' determination to carry out these improvements shows in their pursuit of funding from any available source. A meeting with Hugh Mackay Matheson, Sir James's Commissioner in London in September 1878 gave them enough encouragement to approach the Public Works Loan Commissioners for a loan of around £12,000. The PWL Commissioners regretted it was too late that year but suggested they apply again the following year. If granted, the loan would be over twenty years at 5% interest. The Commissioners therefore approached Messrs Henry Salter & Sons, Glasgow, to see if they could arrange a loan of £7000. When that failed they invited two fishcurers to their meeting on 11 April 1879, Andrew Gibson and James McCombie, and sought their advice on how other harbours had obtained financial assistance.

In the meantime Sir James Matheson died and they were faced with the Trustees of his estate requiring payment of the balance owing on his loan. William Mackay chaired the meeting in which his own letter as Chamberlain was read, demanding repayment of £1000 out of their surplus at credit in the Bank in order to pay towards the loan. The repayment was ordered.

The Commissioners, however, pressed on with their schemes. They increased the Harbour Dues (taking Peterhead Harbour Dues as a model), instigated the 1881 Harbour Order which enlarged their borrowing powers to £20,000 and continued to chivvy the Public Works Loan Commissioners. At a Harbour Commission meeting in September they approved a plan by Mr Kinnear of the extension to the timber wharf and instructed Messrs Mackenzie and Macdonald to prepare Plans and Sections of it in association with the Harbour Master and his Deputy and also in consultation with the captains of steamers frequenting the harbour and other experienced shipmasters. Finally, in February 1882, the Life Association for Scotland came up with the goods in the form of a loan for £10,000 at 4½%.

All this time other small improvements were taking place. A small beacon was erected on Parson's Rock and George Macdonald, Blacksmith, installed a new perch on the Big Rock in Sandwick Bay for £28. On Aeneas Mackenzie's suggestion the Commission offered to pay £10 per annum to the Burgh Commissioners if they employed a scavenger, horse and dust cart to cart away once every lawful morning and evening all ashes etc of the houses and shops in Point Street, South Beach Street and North Beach and the lanes of these respectively. Another £20 was subscribed towards the decoration and cleaning of the Public Quay on the visit of the Duke of Edinburgh to inspect the Royal Naval Volunteers.

A new dredger was ordered from Messrs Priestman Brothers, Hull, at a cost of £389 and the Harbour Master was instructed to provide his assistants with official caps with the initials HMA (Harbour Master's Assistant) on them as well as brass buttons with anchors on their jackets 'so as to distinguish them as being in authority'. By this time there was a harbour boat, a full time clerk in the Harbour Master's office earning £15 per annum and Mr Lees' salary had gone up to £110.

Also at this time it was agreed to contribute one third of the cost of the proposed breastwall at Bayhead up to £150 to 'prevent rubbish being thrown into the river and carried into the harbour thereby filling it up'.

STORNOWAY HARBOUR

"The Petroleum Act 1871"

BYE-LAWS.

In pursuance of, and in exercise of the powers and authorities vested in them under and by virtue of "The Petroleum Act 1871," the Stornoway Pier and Harbour Commissioners do hereby make and ordain the following Bye-Laws, viz.:

I. The Owner or Master, or other person in charge of any ship carrying Petroleum, to which the said Act applies, entering within the limits of the Harbour of Stornoway, as defined in the Act 28th and 29th Victoria, cap. 76, shall moor said ship in the Bay of Glumack, or Glomack, at the south side of the said Harbour, and at such part of the said Bay as the Harbourmaster may direct.

II. If any Ship carrying Petroleum to which the said Act applies requires to land or discharge any portion thereof for repairs or otherwise, the same shall be landed or transhipped under the immediate superintendence of the Harbourmaster, or other person appointed by him; and the hours for such landing or transhipment shall be between 8 o'clock A. M. and 4 o'clock P. M.

The BOARD OF TRADE hereby signify their confirmation of the above Bye-Laws.

By order of the Board of Trade, the 16th day of January 1872.

(Signed) C. CECIL TREVOR,
Assistant Secretary.

Published by order of the Stornoway Pier and Harbour Commissioners.

Wm. ROSS,
Secretary.

Stornoway, 26th January 1872.

EXCERPT from the Act of Parliament 10th Victoria Cap. 27, incorporated with the Stornoway Harbour Order Act, as regards PAYMENT OF QUAY DUES OR RATES.

By Section 42 it is enacted that "The Rates payable to the Harbour Commissioners, in respect of any goods shipped or unshipped within the limits of the Harbour, Dock, or Pier, shall be paid as follows: that is to say if such goods are to be shipped **they shall be paid before shipment,** or if such goods are to be unshipped, **they shall be paid before the removal of the goods** from the premises of the Commissioners, and before the expiration of two months next after they were unshipped."

By Section 43 it is enacted that "If the Master of any vessel or the Owner of any goods **evade** the payment of the rates payable to the Commissioners in respect of such vessel or goods, or any part thereof, **he shall pay** to them **three times** the amount of the rates of which he shall so have evaded payment."

Section 45 enacts that "If default be made in the payment of the rates payable in respect of any such goods, the Collector of Rates may **distrain or arrest** of his own authority such goods, and for that purpose may enter any vessel within the limits of the Harbour, Dock, or Pier in which the goods may be, with such assistance as he shall deem necessary; or if the said goods have been removed without payment of such rates, he may **distrain or arrest any other goods** within the limits of the Harbour, Dock, or Pier, or the premises of the Commissioners, belonging to the person liable to pay such rates, **and may sell** the goods so distrained or arrested, and out of the proceeds of such sale, pay the rates due."

By order of the Commissioners,

Wm. ROSS,
Secretary.

Stornoway,
December 1, 1884.

No 1 Pier

The work on the extension to the timber wharf began in the autumn of 1882. This pier, together with the original structure, has become known to generations of Stornowegians as 'Number One'.

The Commission had employed James Barron of Buckie as their Civil Engineer on the project and during the summer he had taken soundings and borings and examined the Parliamentary plans. He recommended a timber structure instead of a concrete one. Concrete would require dredging down to a hard foundation and would also be 30% dearer. The existing ground, he said, was well adapted for piles which would require to be backed by loose rubble to make up the ground to be enclosed.

Mr Barron was charged with making the best deal he could for the supply of timber and in August he wrote advising that he had purchased 'a cargo of greenheart afloat' on the *Johannes Koster* which was shortly to arrive at Stornoway. The cargo consisted of about 340 logs and the price would be £7 7s 6d per load of 50 cubic feet calliper measurement. The Arrivals Book shows her in the harbour on 18th September 1882 under the command of Captain Schluck with a cargo of timber from Demerera.

There was obviously some friction between the Captain and the harbour authorities. Writing to him three days after his arrival William Ross remarks, 'I think you might with safety have come nearer the shore with your ship which would have facilitated the discharge of the cargo' and the Cash Book records a payment of £6 19s 6d on 17th October for 'the wages of two extra men discharging cargo in respect of his departing from claim for demurrage'.

The weather added to their problems. A letter dated 7th October 1882 from the Clerk to the Timber Merchants in Liverpool said:

> 'I am not able to say how much timber in cubic feet was landed from the *Johannes Koster* before she was driven from her moorings in the recent gale but both the Captain and the Commissioners' Inspector agree in stating it at one half the cargo as near as possible.'

In a note on another letter he states that the vessel is now hauled up on the Patent Slip.

Photographs of the aftermath of this great gale show the barque *Ellen* driven right up on to South Beach. It is not clear whether the cargo of timber strewn around her is her own or that of the *Johannes Koster*, destined for the new pier.

The Commissioners had advertised by hand bills and in the *Inverness Courier* for 8000 tons of rubble but they considered the two responses received too high. They instructed the Clerk to intimate by hand bills that they were willing to receive any quantity of rubble for 3/6 per ton.

Finally in October 1882 they accepted the offer of William Muir, Contractor, Motherwell, to erect a wooden wharf at 7½d per linear foot for driving the piles and 4½d per square yard for planking, materials to be supplied by the Commissioners. Mr Muir supplied an engine and piling machine and said he would complete the work on or before 1st of May 1883, wind and weather permitting.

Commissioner Matthew Russell had sourced a supply of White's Best Portland Cement from James Shaw, Jamaica Street, Glasgow, 'free on board the steamer' at 39/- per ton. They agreed to purchase 60 or 70 tons.

Under way again

The new timber wharf was completed in 1883 after various delays and hitches. The contractor complained that the pitch pine timber promised by Aeneas Mackenzie for the previous November had still not arrived by January; he complained that the piles were bigger than contracted for and were damaging his machine; the *Johannes Koster* still had about half a dozen logs belonging to the Commission on board as ballast as she sat on Aeneas' slipway; Alex Campbell, their Inspector of Works, had moved on and they had to find a new man, and the carters were put out when their payment for stones was cut by 1/- per ton. 'The payment already made is fair and reasonable and the carters ought to be satisfied with it,' said the letter from the Clerk to the Laxdale carter who had complained.

The Harbour Master, Mr Lees, was instructed to see that in future no horses and carts were permitted on any account to go on any part of either of the timber wharves, which was easier said than done. 'Some of the Commissioners,' the Clerk told him, 'have again been complaining. Please let me have the name or names of any person or persons whom you find infringing your orders in this respect so that we can take the Commissioners instructions on the matter.'

One of the persons who infringed the orders – in fact drove a horse and cart through them – was Andrew Gibson, fishcurer. Mr Gibson had been delegated by the Herring Smoking Curers to address the Harbour Commissioners at their meeting on 30th April 1884. The curers were finding that the iron posts erected along the quay to stop access by horses and carts put them to a great deal of expense when shipping their fish. They asked that they be removed and their horses and carts allowed on the wooden wharf.

The Commissioners were sympathetic but pointed out that the wood got cut up very badly by such traffic. As a compromise they offered to supply moveable platforms in connection with the steamers taking in smoked herrings, provided the curers signed a written obligation to strictly confine the horses and carts to such platforms.

Mr Gibson's response is recorded in a letter, dated two days later, from the Harbour Master, who wrote:

> 'I went and ordered the carters to take their carts off the wharf but Mr Gibson immediately ordered it on again. I then put on the chain and padlock – Mr Gibson took a hammer and broke the iron which attached the chain to the posts – so that unless the Regulation is enforced by the Law it will be impossible to keep horses and carts off.'

Fighting for the Foreshores again

The year 1884 saw the first recorded exchange in what was to become a running battle between the Stornoway Pier and Harbour Commissioners and Lady Matheson, the widow of the late proprietor.

Lady Matheson

When Sir James Matheson had purchased the foreshores of Stornoway, he had the environmental improvement of the town – and the Castle policies in particular – more in view than the potential for the fishing industry. He had already made his fortune in the China seas, through the opium and tea trades and with estates on both sides of the Minch, he was now the largest landowner in Ross-shire. He did not need, therefore, to make a commercial success of the port.

However, having accepted that the feuars and merchants of Stornoway were determined to have a say in matters, he seems to have applied his business mind to their deliberations and encouraged and financed any scheme he thought worth while.

Not so Lady Matheson. Twenty years younger than her husband, Mary Jane Perceval was born in Somerset, one of several daughters of Michael Henry Perceval of Spencer Wood, Quebec, Canada. At some time her father was a member of the Legislative Council of Quebec. She arrived in Lewis very shortly after her marriage and was Lady of the island for fifty-two years. The couple carried out many improvements to the island, using Sir James's considerable wealth. Roads, bridges and

Peter Bachelor

William Mackay, chaimberlain of Lewis, 1875 – 1889

C Scott Mackenzie

William Ross, first Clerk to the Harbour Commissioners

schools were constructed as well as the new Lews Castle and its extensive grounds. Lady Matheson felt very strongly that the Matheson ownership had improved the lot of the islanders greatly.

As proprietrix, she was entitled to take the chair at Commission meetings but, whether from choice or convention, her Chamberlain stood in for her. However, all major decisions required her consent. From her letters to the Commission and from press reports of the time, she comes across as a strong character with very definite political views and a sometimes querulous nature. Described by Matthew Russell as a 'kind and amiable lady' and given to acts of generosity, she was also very conscious of her position. Her view was that she was holding the island in trust for Sir James' heirs.

Sir James had died just as the question of land reform had begun to come to a head. During the 1880s the rising demands for land and franchise among the Highland population began to alarm the authorites. In the autumn of 1884 demonstrations were taking place in such law-abiding places as Peterhead and Nairn, with thousands marching and listening to speakers calling for radical change. In Lewis William Mackay was making a second attempt, in his capacity as Chamberlain, to collect rents from angry crofters in Uig; a large Land Reform demonstration took place in Stornoway in November and

MacBrayne's steamer *Lochiel* was kept in port at Stornoway in readiness to pick up a contingent of policemen in Strome and transport them to Skye, should any trouble flare up there. The press carried vivid reports of disturbances and long letters of grievances to the editors by the leaders of the land reform movement.

It was against this tense background that the irresistible force of the Pier and Harbour Commissioners met the immoveable object in the shape of Lady Matheson.

Now that the new wharf was operational the Commissioners were considering further expansion. They had decided to erect new storage accommodation on the quay and Lady Matheson did not object, beyond commenting that she hoped that the new sheds would be 'more ornamental' than the present ones.

However, one of the newer Commissioners, Captain Thomas Morison, had suggested that they apply to her Ladyship for a feu of the foreshore between the Patent Slip and Goat Island (then sometimes referred to as Coul Island) which they could infill by degrees with harbour dredgings and so provide new areas for curing stations. Lady Matheson's letter of response is not minuted but was negative.

In a letter to the Engineer in May 1884 the Clerk refers to her objections and asks for Mr Barron's views on the proposal so that they can approach her again on the subject. He adds:

> 'I may mention that forty-five years ago Coul Island was one Island and the ground in the sound was higher than it is at present. Now you will observe that the Island consists of two portions owing to the action of the sea.'

They tried again in September and she replied, saying she had fully thought over the application and she was still of the same mind, 'that I do not think it advisable to feu these foreshores, if the ground when reclaimed is to be used for curing stations, for curing there would be as objectionable as on the South Beach.' She was quite prepared to remove the curing stations from South Beach and provide new stations behind or north of the Smoking Houses. She would also share with the Harbour Trust and the Road Trustees the cost of a retaining wall along Newton. If all that was done. . .

> 'I do not think there would be any necessity for the Harbour Trust to fill up and reclaim the opening between Newton and the Island, the cost of which would be very large and much more than the Harbour Trust with their present amount of debts could well afford to meet.'

Captain Morison, who had first voiced the idea, moved that they appoint a deputation to wait upon Lady Matheson and explain matters to her 'in the hope that she may be induced to change her mind on the subject'. Lady Matheson sent word by the Chairman that she did not think a meeting was necessary.

The members of the intended deputation – Matthew Russell, Kenneth Smith, Murdo Macleod and Thomas Morison – then framed a Minute to send to her on the subject. The document, addressing her Ladyship's objections one by one, paints a picture of an industry fast outgrowing its accommodation.

> 'Every season makes it more evident that the whole of the South Beach is required for foot and vehicular traffic, which has grown enormously since the establishment of kippering premises – it being the only thoroughfare used between these premises and the public Wharf and Quays.'

Transferring the South Beach curing stations to north of the Smoking House would mean much loss and deterioration in the quality of the fish. It was 'absolutely necessary' to dredge the North Harbour and the area around the wooden wharf as steamers were already touching the bottom there and where were the dredgings to go if not into the proposed reclamation?

The document was despatched to the Castle with a note from the Secretary, explaining that he sent it in terms of his instructions from the Committee who also required that it should become part of the Commission's Minute. He signed it, perhaps in some trepidation, 'I am respectfully, Dear Lady Matheson, Yours truly William Ross, Secretary'.

The opening paragraph stated the Commissioners felt that probably 'Lady Matheson has not fully considered all the interests involved in this question.' In fairness to her Ladyship, she sent word that she was reconsidering.

In November William Mackay gave the Commissioners a list of her objections and the Committee answered these immediately. This time William Ross told her that the deepening of the harbour was urgently needed and the Commissioners were anxious to proceed. He ended, 'Your Ladyship will no doubt be gratified to learn that Revenue of the Harbour during the year ended yesterday is more than double that of any previous year.'

Lady Matheson was unconvinced. In December she again said 'No' in a long and frequently underlined letter, suggesting among other things that the dredgings could be used by farmers as top dressing. There was no more valuable top dressing for land and grass, she stated, than the mud of a harbour.

Highland News

In May 1887 the *Highland News* reported that the entire fishing fleet, amounting to around 500 boats, landed around 10,000 crans of herring in one week at Stornoway. The report went on:

> 'Owing to the number of boats the berthing space is very scant, and crews exert themselves to gain a point of vantage from where to land their catch with as little delay as possible. It is pleasing to note that, notwithstanding the bustle and confusion occasioned by such a large number of boats eagerly working windward in a narrow space, accidents are very rare.
>
> The fish-carrying steamers have been kept busily employed. The SS *Eleanor* made daily trips to Stromeferry until Saturday when her steering gear gave way and she had to put back to have it repaired. The steamer *Clansman* made a special run to Stromeferry with the *Eleanor's* cargo on Sunday morning. The steamers of the Stornoway Fish Carrying Co were all so fully occupied during the week that the Company had under consideration the advisability of placing additional boats on the route. . .
>
> Some idea of the extent of the kippering trade directed to Fleetwood by the Company can be gathered from the circumstance that over 60,000 packages exclusive of fresh fish have already been landed there since the season opened, the steamers *Diamond, Sapphire* and *Ben Nevis* and *Pearl* (sic) now being on the route.' ❑

By now tempers were fraying. The Commissioners sent word that they were unanimously of the opinion that the Newton foreshores should be feued, as proposed, in the public interest and hoped that her Ladyship would, without further delay, consent to approach Sir James' Trustees or allow the Commissioners themselves to do so.

'As to Farmers using dredgings,' they snapped 'the Commissioners wish to point out to Lady Matheson, that they have already tried this plan by placing a considerable quantity of dredgings on the quay but the farmers failed to cart the stuff away and it had consequently to be removed at the expense of the Burgh Commission.'

Finally in February 1885 William Mackay brought Round One to a close in a letter from the Chamberlain's Office:

> 'A few days ago I had an interview with Lady Matheson when she gave me to understand that she looked upon her last communication to the Pier and Harbour Commissioners as final in regard to feuing the Newton Foreshore. The reasons put forward by the Harbour Commissioners for obtaining these Foreshores and the use they purposed making of them has not convinced her Ladyship it would be any advantage or improvement to the Town or Harbour but the reverse and might materially interfere with the rights of the Estate in the future. . .'

If ever there was a lady not for turning, it was Mary Jane Matheson.

The phrase in William Mackay's letter regarding the rights of the estate sums up Lady Matheson's entrenched attitude to change. Her ordered world was under siege and she had lost her husband. Her fear of losing anything of what he had built up shows from time to time, when she describes herself as holding the estate in trust for his successors.

The Commissioners retired from the fray but only temporarily. They had plenty to keep them occupied over the next two years, beginning with the necessary dredging work. The Harbour Master received instructions that it should be started immediately, working inwards from the outer end of the wooden wharf and depositing the dredgings at the old Distillery Quay which, according to the Minute, Lady Matheson had 'already sanctioned'. The Commissioners had advertised for a man and 'a stout active boy of about fourteen years of age' to take charge of and work the dredger barge. Phillip Macleod, Sandwick, offered 18s per week to do the job and was accepted, but whether the price included a stout boy is unclear.

By this time William Lees had relinquished the duties of Harbour Master, although he remained Collector of Dues. The Commissioners, perhaps aware that the combined post was too much for one man, had appointed one of their own number, Captain Thomas Morrison, as Harbour Master. At fifty he was recently retired from a career begun in the China trade and ended on the Atlantic routes between Britain, Newfoundland, Spain and South America, carrying salt, dried fish and wine. On his return to Stornoway he had immediately pitched into public life, becoming a Burgh Commissioner, and later a Harbour Commissioner. He was obliged to give up his seat on the Harbour Board after his appointment but remained on the Burgh Commission.

Thomas Morrison was to be Harbour Master for twenty eight years, retiring in 1913 by which time the port would have changed almost beyond recognition. At the time of his appointment however, he was faced with the problem of too many vessels looking for berths and not enough to give them.

It was obvious that something had to be done about new quay space. In August 1887 the Commission received a petition signed by a large number of fishermen from various ports in Scotland, complaining about the inadequate accommodation at Stornoway 'in consequence of the extra pressure caused by the method of disposing of fish by auction recently adopted.'

A month later the Burgh Commissioners added their support to the campaign, calling on the Pier and Harbour Commissioners to:

> '. . . consider the deficient harbour accommodation whereby loss results to the fishing interests of this Port and the burgh generally, with a view of providing as soon as possible sufficient quay and fish curing accommodation which the (Burgh) Commissioners believe can best be obtained by extending southwards the present South Beach retaining wall and filling in the reclaimed space'.

In the autumn, having received the approval of the Board of Trade, they moved things a step further by asking Alex Macdonald to prepare plans for the proposed new quay on South Beach which they included in the draft of a new Provisional Order. They also instructed that their borrowing powers authorised in the proposed Order should be increased to £50,000 to allow them freedom to complete the necessary improvements.

This, of course, opened the second front of their battle. William Ross wrote tentatively to the Chamberlain, William Mackay, on 24th November 1888:

> 'Now that you have got relief from the arduous duties you have for sometime been engaged in, may I ask if something has been done in the way of preparing a feu charter by Lady Matheson in favour of the Commissioners and if not how soon the draft can be submitted for revisal by the Commissioners?'

Mr Mackay's duties had, indeed, been arduous. The Lewis land struggle between the increasingly angry and desperate crofting population and the doggedly resistant Lady Matheson had continued over these last three years, culminating in January 1888 in the famous 'Aignish Riot'. The changes within the fishing industry and the demands, not only for quay space, but also land for curing and kippering operations, gave him extra problems. He was chairman of several local bodies, each with its own particular agenda in a fast changing society. Now over sixty, it was rumoured he had wanted to retire two years earlier but Lady Matheson wanted him to stay on to take the estate through the challenges of the crofters in the Land Court. He had travelled the island that autumn, representing the estate at the sitting of each Land Court in each district, giving evidence and examining witnesses

His employer, in the meantime, had ventured a little into the field of politics by inaugurating two branches of the Primrose League, one in Stornoway, the other in Uig. In this Lady Matheson had been assisted by an otherwise unidentified 'Hon Miss Murray' and the newspaper report on the event described it as 'falling like a thunderbolt on the political portion of the community'.

The Primrose League was a Conservative organisation founded by Lord Randolph Churchill and John Gorst with the aim of helping the Conservative Party adapt to the extension of democracy, which leading Tory figures feared would be a challenge to authority. At the time of Lady Matheson's involvement its members were around 200,000 and it would grow to over one million in the following five years. By the end of World War I it had become absorbed into the official party structure but at the time of Lady Matheson's involvement it was on the rise.

At the opening of both of these local branches in 1886 her Ladyship had pictured 'in glowing indignation, the deplorable state of the country, the evident downward progress of the star of the British Empire and the atheistical aggression upon our Constitution'. In an interview for the Glasgow *Evening Citizen* a week or so after the Aignish disturbances, she remarked that the real explanation for the misrepresentations she had been subjected to in the press was that:

> 'I have given tremendous umbrage to the Land League by speaking about them at my little meeting here of the Primrose League. I said they did not respect the ordinances of the Eighth Commandment and that their agents did their best to prevent to foment evil between the landlords and the people. . . I do what I can to help the people but they will not help themselves when they might do so. . .'

There must have been times when the Chamberlain was sorely tried by Lady Matheson. One of her critics, Donald Macrae, Balallan schoolmaster, described her as pretty, 65 years old and spending all her time lecturing her gamekeepers on the principles of the Primrose League.

Whether by accident or design William Mackay seems to have omitted to tell her much about the plans of the Pier and Harbour Commissioners to do something to help themselves. By the end of 1888 they had revised their proposed Provisional Order to give them £70,000 borrowing powers as they realised the works were likely to cost over £40,000. They had the plans, the approval of the Board of Trade and all they needed was the feu charter from the proprietrix.

Lady Matheson objected to the Board of Trade at the start of 1889:

> 'I assume the Board of Trade are aware that the foreshores in the Harbour of Stornoway are my property as Life Renter and I have the following objection to urge as well as certain conditions to make before I can agree to the proposed Order.'

Her objections were that the Order asked for ten years to complete the works whereas the Board of Trade would only sanction five years. She feared that this was utterly inadequate. She would not give them all of the foreshore they asked from Newton to the island but would agree to South Beach subject to conditions. These were three in number – that the three intended wharves be constructed of open work of iron rather than solid masonry in order to prevent silting; that no other buildings or works be erected on the foreshores; and that no 'Fish Stances' be permitted on any part of these quays.

The Commissioners replied that it was perfectly possible that the works would be completed within the stipulated five years and if not there would be little difficulty in carrying through a brief Order for an extension of time. They desired to impress upon Lady Matheson:

> 'as forcibly as they can the absolute necessity of carrying out the proposed works without delay. . . Looking to the fact that the present Harbour accommodation is inadequate to meet the requirements of the ever increasing fishing industry of the district; looking also to the proposed Harbour extension at Thurso and to the imminent probability the establishment of fishing ports on the West Coast of Sutherlandshire unless traders are afforded all available facilities at Stornoway, they fear that Lady Matheson's refusal to grant the ground required for the proposed extension of the accommodation at Stornoway Harbour will lead to the most disastrous results. . .'

They regretted that there should be a difference of opinion on a matter of such great importance to the district as well as to the superior. They said they had never intended to construct quays of solid masonry but of greenheart timber and had no objection to her Ladyship's suggestion of iron pillars, should these prove as cheap and durable. They fully agreed with Lady Matheson that the fishcuring operations should be removed from the vicinity of dwelling houses which was why they proposed to extend the South Beach quay wall, thus permitting 'a thoroughfare of about sixty feet being left between the pavement on South Beach and any curing stances that may be necessary'.

Within a week her answer came back, taking their points in order and underlining occasionally. Starting with the Commissioner's view that an extension of time would easily be granted she said:

> ' I do not know who their Informant is but, I can assure them, he is mistaken and there is no 'matter of course' in the case, for as a matter of fact the 'extension of time' would depend on the report of the Board of Trade official who would be sent to inspect the work done . . . should the report happen to be unfavourable the 'extension of time' would be refused without doubt.'

Moving on to the positioning of curing stances she said her objection

> '. . . was not so much the want of room. . . but the very great and extreme nuisance caused by carrying on Fishcuring in such a very prominent and conspicuous position. . . The system is unsanitary to the last degree especially during the warm weather; moreover it is so unsightly and forbidding that it should not be allowed in the open air at all but should be entirely carried out in covered yards removed from the houses and harbour as at present done by most of the Fishcurers'.

As regards the construction of the quays she remarked that she believed there could be:

> 'no question as to the superior durability of iron over greenheart; as to their relative cheapness I can say nothing; there are frequently very heavy seas on the South Beach so I should recommend iron.'

Finally she gave vent to her annoyance at being kept in the dark.

> 'It is to be regretted that the Commissioners did not acquaint me from the first with their wishes on this whole matter for then our mutual conditions could have been arranged before applying for the Provisional Order; in similar cases I am informed this course is almost invariably adopted, and thus, not only is unnecessary expense obviated, but delay and misunderstanding avoided.
>
> I knew nothing of the Order at all until I saw it advertised in the public papers. . . In conclusion I beg to assure the Pier and Harbour Commissioners that I take the warmest interest in everything

> connected with our fishing industry and the conditions I now attach to these feus are entirely framed, in my opinion, for its benefit.'

It is difficult to believe that it would have slipped the Chamberlain's mind to inform his superior of such a major undertaking as a Parliamentary Order changing the face of the entire harbour, even with all the disturbances and political agitation in which they were both involved. Perhaps he had told her and Lady Matheson had not grasped the full import of it. Perhaps he deliberately left the matter to resolve itself without bringing more tirades about his head. Perhaps he let the Commissioners have enough time to organise themselves before taking her Ladyship into the matter.

Certainly the Commissioners, recording their profound regret at his retiral in May that year, expressed their appreciation of 'the active interest he has always taken in the improvement of Harbour accommodation and generally in the trade and interests of this Port.'

Provost Roderick Smith, writing in the following century, credited him with being 'shrewd, intelligent and practical in his business methods' which would suggest that he would have handled the foreshore battle with an eye to the business potential of the port.

Now in poor health, he took his doctor's advice to retire and left with the young wife he had married only three years before, for a more peaceful life on the mainland. In recording his retiral the *Highland News*, no friend of estate owners, described him as graduating at the feet of Donald Munro in the secrets of Highland estate management. While admitting he was of a very different character to his tyrannical predecessor, the paper still saw him as the Matheson's enforcement agent against the people. Munro did his work with undisguised open force, it said, while Mr Mackay used 'miserable estate-craft'.

On leaving the island, however, William Mackay was presented with a Public Address by members of various public bodies, including the Harbour Commission, all of them testifying to his calmness, sound judgement and courtesy in difficult and trying times. John Ross, enclosing an Extract of the Minutes of a Burgh Commissioners' meeting, added a personal note concurring with their sentiments and speaking of the 'pleasantness that always characterised our frequent meetings together. Lady Matheson gave him a large silver tray, inscribed with her grateful thanks for his faithful service as Chamberlain. William Mackay died in 1901, at his home town of Nairn, aged 74 years.

The last push

Thirty-eight year old William Maclennan, assistant in the Estate Office, now took over as Chamberlain and Chairman of the Harbour Commission. Two weeks after their last communication the Commissioners noted that, although her Ladyship had not yet agreed to give them the ground for the proposed new works, she was willing to let them have the ground in the Inner Harbour for £7 feu duty an acre and on condition there would be no curing. They therefore instructed the Clerk to telegraph acceptance of these terms to her legal agents and waited.

Two years went by, during which they continued their quest for funds and their immediate improvement plans. In the summer of 1889, it was agreed that the position of Secretary/Treasurer should be shared by William Ross and his son John at a total salary of £80 per annum. The Rosses had accepted a fixed annual salary rather than make professional charges for the extensive law work they had been undertaking for the Commission.

The congestion at the quay by the 1891 season was such that three more berthing masters were employed at 25s per week and the Harbour Master was instructed to enforce the regulation compelling boats to leave the quays after discharge of their herring.

Aeneas Mackenzie, boatbuilder, shipowner, and proprietor of the Patent Slipway on South Beach, rejoined the Commission that year. A stubborn and fiery individual, he remains, like John Norrie Anderson, a legendary figure in the history of the town. One of Aeneas' first actions as a Commissioner that term was to move that two new wharves be built on South Beach for the landing of fish, one to be at the foot of James Street dealing with the kipper traffic and the other at the foot of Cromwell Street, specifically for Baltic shipments. The motion was seconded by John M Morison and unanimously carried.

The Secretaries wrote to Lady Matheson at her London residence on 13th July 1891, asking her to grant the Commission the ground for the construction of those two piers, which they believed would meet the immediate necessity. However, they reminded her that they still considered that the whole of the ground in the South Beach and Newton area would be required sooner or later. They appear to have had some encouragement in a letter from her unrecorded in the Minutes but on the strength of which they ordered a cargo of timber for the proposed new wharves.

It had been three years since Lady Matheson had been in residence at the Castle and she arrived from Ullapool on board Sir James Fowler's yacht on the last Saturday in August 1891. She was met by the officials of the town, as well as a detachment of the Volunteers, and hundreds of the islanders turned out to watch. Two weeks later Kenneth Smith, Murdo Macleod and Aeneas Mackenzie reported to the other Commissioners on a meeting they had had with Lady Matheson at the Castle in which they had 'laid before her Ladyship as fully as they could the present position of the Commission financially and otherwise and the necessity that exists for an immediate extension of the quay accommodation if disastrous results to the traffic of the port and the interests of the community are to be averted.'

They said that she had granted them a patient and courteous hearing before presenting her own views which had apparently been arrived at 'after long and painstaking inquiry and consideration of the whole circumstances'.

These views were that no further extension of harbour accommodation should be attempted until the harbour debt of £16,000 was wiped off or at least considerably reduced and until they had some knowledge of the effect a railway from Ullapool or other mainland point would have on the fishing industry, perhaps necessitating a much larger extension than the Commissioners had in view. Beyond that Lady Matheson had made no further comment. She had not given a decided refusal to feu the South Beach foreshore or part of it 'which consent is a necessary preliminary to all further negotiation on the subject'.

In the Chair at that meeting was the new Chamberlain, William Gillanders, who had been Estate Accountant for over twenty years before taking over from William Maclennan, who had moved on to pastures new as factor of Preston Hall Estate in Dalkeith. Mr Gillanders found himself in the same position of go-between as William Mackay had some three years before but while Mackay seems to have given tacit approval to the Commissioners' plans, Gillanders seems to have been more involved in the decision making. Certainly the Minutes of the next few stormy meetings show changes to the wording, initialled by him, some phrases being crossed out and others added in or re-written.

Along with Commissioners Murdo Morison and John M Morison he considered the report of the deputation and signed the Minute which stated that 'the express opinion of the meeting was that the proposed extension of quay space in South Beach would brook of no delay except at great sacrifice to the public weal' and made reference to Sir James Matheson's desire to develop the harbour, his shrewd forethought and his handsome loan which had enabled the first extensions to quay space. The present debt was very moderate as compared with that of other harbours of less importance than Stornoway and was represented by nearly three times its amount in new and permanent property and plant.

This Minute was sent to her Ladyship with the request to reconsider and especially not to overlook the consequences which absolute refusal might have on the inhabitants of Lewis with whose true welfare and prosperity the Commissioners had not the slightest doubt Lady Matheson had the fullest sympathy and anxious concern. It would be interesting to know who framed the wording of it

William Gillanders submitted his memorandum on Lady Matheson's views to the next meeting and hoped that the Commissioners would 'see the generous manner in which her Ladyship views the necessities of the situation'.

This time she again stressed her great disinclination to get into debt and the importance of ascertaining what effect a mainland railway terminus would have on the trade of the port. Again she stressed that she believed the Commissioners should wait for the report of the Railway Committee which was looking into such a provision, as well as on what terms the required funds would be raised. Then she asked for written figures to show how the Commissioners had arrived at their assertion regarding the proposed expenditure bringing in more revenue, asking what would happen if the numbers of visiting boats were affected by increased or a failure of the fishing.

She said there was no further point in discussion until the Commissioners' application to the Treasury had been dealt with one way or the other and in order to get the application in as soon as possible she offered to pay the costs of the Provisional Order, as she had done for Carloway.

The Memorandum ended:

> 'The Commissioners should be aware that Lady Matheson has no other interest in this matter than what is for the welfare of Stornoway, but, as Superior, the privilege should be allowed her of submitting to the Commissioners what in her opinion is the best mode of carrying out the extended scheme proposed. Lady Matheson will offer no objection or opposition to the Provisional Order and had never contemplated doing so.'

The Commissioners recorded their satisfaction that her Ladyship would now offer no objection to the Provisional Order and, while thanking her for her generous offer to pay the costs, hoped they would not have to avail themselves of it since they had always contemplated the necessity of it and believed it should be obtained at the Commission's expense.

Victory

In the end, that old enemy time won the battle. It ran out before the Commissioners could submit all the required documentation for the Act of Parliament and John Mackenzie, who had predicted that time would tell, gave notice that he would move that the Provisional Order be proceeded with. John Ross, by this time sole clerk to the Commissioners, wrote the Parliamentary Agents to prepare them, explaining events over the past few weeks and the 'considerable amount of irritation displayed by the Commissioners and the public generally. . .This accounts in some measure for the somewhat contradictory instructions I had to forward.'

He was referring to a fiery Ratepayers meeting at which a resolution had been passed approving the idea of attempting to obtain compulsory purchase powers for the foreshore. A short sharp contribution to the debate that night had been made by Colin George Mackenzie, Solicitor, in which he praised the patience of the Commissioners. He predicted that Lady Matheson's name 'would not go down to posterity under the imperishable claim which ought to be the aim of every true woman but as one that paralysed and crippled trade, especially in the kipper department of Stornoway which was the envy of the civilised world.' He had been loudly cheered.

Somewhere along the way, an offer had come from one John Otto MacQueen, Advocate, 22 Bridge Street, Aberdeen, to mitigate between Lady Matheson and the Commission. They had delayed taking up his offer during their own internal arguments but on 15th December 1891 he and John Ross met Lady Matheson's legal representatives in Edinburgh and put proposals for a partial settlement to them, asking for some of the ground immediately required for quays to be granted.

Messrs Stuart and Stuart responded in January 1892 saying that her Ladyship had given the matter careful and anxious consideration. They suggested that it was to the Commissioners who were perpetuating the irritation between the proprietor and the public, which they said Lady Matheson had done nothing to call forth. She may have differed in her views from the Commissioners but she wholly repudiated the motives which had been attributed to her in public and had done nothing to justify the attacks made upon her.

John Otto MacQueen met Lady Matheson on 28th January 1892 and reported to the Commissioners that afternoon. The meeting learned that her Ladyship would not exclude curing operations on any part of the ground now in question; that she offered to keep the feu duty at £3 per acre for seven years and then at £6 per acre if required by the Superior; and that the Commission would not be required to pay for the roadway or esplanade between 25 feet from the kerb on South Beach Street and the seawall which would be maintained by Lady Matheson as part of the public thoroughfare and would not be encroached on for curing or any other purposes.

The offer was accepted, those members of the Commissioners who were Burgh representatives stating that their decision was subject to confirmation by their constituents. However, as always in this long-running battle, it was not over yet. Mr MacQueen met the Commissioners again the following day and reported that there had been some misunderstanding between himself and Lady Matheson regarding the restriction of feu duty.

She had intended the restriction to apply to the South Beach ground only, while he understood it to apply to both South Beach and the Inner Harbour, for which the Commission had already agreed to pay £7 an acre. The Chairman read a letter in which she said she felt it was a bad precedent to alter feu duty after it having been agreed but she was willing to apply it from Martinmas 1892 instead of from Martinmas 1889, at £3 for the next four years and £7 from 1896.

With that last salvo the conflict was more or less over. The Commissioners resolved that:

> 'considering the expense and inconvenience involved in further delay in executing the works, it seems necessary now to accept Lady Matheson's terms which has accordingly been done without prejudice to any rights in the foreshores competent to the community of Stornoway.'

They told John Otto MacQueen to ask Lady Matheson's Edinburgh agents for a Draft Feu Charter.

Later in the Spring letters appeared in the *Inverness Courier* and *The Scotsman* signed by William Gillanders and quoting from Harbour Minutes regarding the dispute. When called upon to explain why he had written them, he used the age old excuse – 'She made me do it' – and read out several letters from her to prove it. The Commission directed the Secretary to inform Lady Matheson, through the Chamberlain, that they were not responsible for what the press wrote.

The sad thing is that Colin G Mackenzie's prediction came true, in a measure. While Lady Matheson did not destroy the trade of the town, or even set out to do so, she certainly delayed its expansion. It is likely that if her husband, whose business acumen is credited with establishing the port of Hong Kong, had lived longer, the necessary quay accommodation would have materialised much earlier. Despite all the benefits she gave the island, despite all the little individual acts of kindness, and despite her desire to protect the interests of the estate and the island, she is remembered more as a hindrance than a help to the community.

John Maclean

The forest of masts in the Inner harbour highlights the congestion at the quays in the early 1890s

Herring girls pose for the camera (top), their cheerful faces belying the long hours of backbreaking work during the fishing season

both photos – Western Isles Libraries

At the Court of King Herring 1892–1913 3

IN THE SPRING of 1889 fifty families from Lewis, Harris and Uist left Stornoway on the steamer *Claymore* on the first leg of the long journey to Canada. They were mostly young, some of them married couples with small families, the children bewildered by the noise of the steamer's whistle and the shouted messages between those on the ship and the crowds on the quay. Three of the couples had only been married a matter of hours before sailing. The Government Emigration Scheme promised them a better life across the ocean.

Although the town of Stornoway was entering on possibly the busiest and most economically viable period in its history through the growth of the herring industry, the rural communities were still struggling. The herring fishing did bring money into the villages from the men and women employed in the boats and at the gutting, curing and packing, but it was seasonal. During the rest of year it was hard to make a living from the land alone and the cry for more crofting land continued. In April 1892 the press was reporting severe poverty in the Barvas area of Lewis, due to the failure of the crops and the severe winter weather and a committee had to be formed to distribute much needed relief.

While the foreshore battle had waged, taking up almost all the time at the Pier and Harbour Commission meetings, other events were taking place in the port and in the wider arena of town and island life.

In September 1887 the Admiralty took a decision to establish a Coastguard station in the island. Joseph Chamberlain MP received a letter from Lord George Hamilton saying that negotiations for the acquisition of quarters were almost complete. Lord George felt that Mr Chamberlain should be the first to hear of the result of one of his own suggestions, since he had taken so much trouble and interest in the condition of Lewis.

In November of the same year the Royal National Lifeboat

Institution established their Stornoway Branch with Aeneas Mackenzie as the Honorary Secretary. The inauguration took place on 7th December and the new lifeboat *Isabella* was put through her paces by the first coxswain, Norman Campbell from Newton, a retired soldier who began his army career during the Crimean War. The Commission (which a decade earlier had felt there was no necessity for a lifeboat in the port) granted the use of the harbour dredger to take part in the demonstration.

In January 1892 the Commissioners made a payment of £3 3s to Mrs Peggy Macleod of the Sailors Home 'in appreciation of her past services in cases of drowning and the inconvenience she has been put to in supplying bedclothes.' The Burgh Commissioners had also given the sum of £1 and the *Highland News* reporting their meeting added that Peggy was justly described as a kind creature.

But the most pressing need at this time was the arrangement of enough funds to carry through the long awaited improvements and in February 1892, on the suggestion of John Otto MacQueen, they sent a deputation consisting of John M Morison, Aeneas Mackenzie and John Ross to meet him in London with the purpose of lobbying as many Members of Parliament as they could. James MacCombie, one of the

C Scott Mackenzie
Peggy

Peggy Macleod

Donations made to Mrs Peggy Macleod by the Pier Commissioners in 1892 were possibly prompted by an incident which took place that winter involving Otto Olsen, a Norwegian seaman, who had been taken ashore from his storm-damaged barque *Nimrod* with a fractured thigh. He was taken to the Sailor's Home which at the time was about the closest thing to a hospital the island.had, apart from the 'Fever Hospital' on the outskirts of the town which dealt with infectious diseases.

The Home had been established by Sir James Matheson and was located in a building, now demolished, which stood on Esplanade Road, next to present-day Amity House. It had one long room with a paved floor and was lined with tiers of bunk beds. In 1892 it was presided over by Peggy Macleod whose formidable appearance in a photograph of the time belies her kindliness and care for the many weary, sick, and sometimes destitute, souls who passed through the door. Any reference to Peggy in press reports or official records is appreciative and affectionate and it is a tribute to her that the local doctor in 1892 had taken to sending his most urgent cases to the Sailor's Home for nursing.

At the time of Otto Olsen's accident, Lady Matheson was in residence at the Castle and, having heard of the event, came to visit him and was impressed with level of care he was receiving. On the very day that Otto left the Home, another fracture case was brought to Peggy, this time one William Brass, who, while helping a fellow passenger unload a trunk, fell between the steamer and the quay and fractured his skull. He was nursed at the Home for a long time and Lady Matheson, seeing the needs and possibilities, set about getting plans prepared for alterations and improvements and brought the facilities at the Home up almost to hospital standards. For the next few years some difficult operations were carried out there and many seriously ill patients treated. ❑

Peterhead fishcurers, was also asked to help. Otto MacQueen believed that if the claims of Stornoway and the island were brought before the Public Works Loan Board in what he termed 'an adequate manner' they could secure a loan of around £30,000 at a favourable rate.

The deputation left for London on the night of 23rd February, travelling straight through. John Ross made the arrangements, writing to the Superintendent of Line at the Highland Railway Company to book first class accommodation and remarking that he supposed they would not get a berth in a sleeper until they reached Perth. 'Perhaps you will kindly wire on receipt hereof what is the best you can do for us and what route you would recommend. Apologising for troubling you,' he ended. His telegram to Cesari at the Station Hotel, Inverness is much more direct. 'Have luncheon basket for three gentlemen passing south at three o'clock tomorrow, Ross, Secretary, Harbour Commission.'

Once settled in the Westminster Palace Hotel in the capital, the deputation set about their business with a will. All strong, articulate characters and all determined that there should be no more delay in this business, they would have been a persuasive trio. They took advantage of the Annual Conference of the National Sea Fisheries Protection Association being in session and, although the matter was not on the agenda, they managed to have Stornoway's needs discussed and to get the unanimous backing of the delegates for their loan application. They then moved on to have interviews with the chief officials of the Board of Trade and the Public Works Loan Board where they were received courteously and sympathetically and given informal assurance of favourable consideration, as long as they complied with necessary regulations. They spoke to the Under-Secretary for Scotland and had his assurance of 'cordial co-operation'. They also roped in the assistance of Dr Macdonald, the County Member of Parliament, as well as that of Mr Anstruther and Mr Fraser-MacIntosh, two Scottish MPs whose knowledge of legal procedures made them valuable allies.

While in London they went to considerable trouble to arrange insurance cover amounting to £15,000 for the dredger *Kyle* which the Commission had agreed to hire from Ayr Harbour Trustees at a cost of £90 per week. Not content with that, hearing that the Ayr Harbour Trustees had been a trifle sticky about letting the dredger sail for Stornoway until they had the actual policies to hand, the three men made a detour to Ayr on their way home on 4th March and arranged that she should sail that night which, says the Minute, she accordingly did. Deputy Harbourmaster Peter Macpherson had been sent down to Ayr to assist in piloting the vessel to Stornoway. She arrived on the night of Sunday 6th March and, by the time the deputation reported back to the other Commissioners two days later, the dredging was proceeding in a satisfactory manner.

The deputation was cautiously optimistic. They were aware that nothing could be guaranteed but they had been encouraged by their reception. 'The main question,' their report stated, 'appears to be whether the works proposed provide such refuge and shelter as will entitle you to an advance of the whole or just part of the money. . . The determination of the question will again twin on the point whether in the main the proposed works will sufficiently provide refuge and shelter or will merely afford additional quay and trading facilities for the benefit of the port itself.'

The Board of Trade officials had pointed out that by omitting the proposed breakwater from Newton to Goat Island from their plans they had departed from the original scheme which gave shelter and refuge. The deputation had replied that they were powerless in this regard since that part of the foreshore had not been part of the agreement with Lady Matheson. They recommended that the Commissioners give very careful consideration to altering the plans sufficiently to provide enough enclosed sheltered harbourage to comply with the Board of Trade requirements.

William Gillanders, chairing the meeting, undertook to ask Lady Matheson for permission to erect a breakwater from Newton to Goat Island. The Commissioners did not expect a favourable response, and John Ross writing to tell John Otto MacQueen of how things had gone on their return, remarked that it was doubtful if she would agree. However, Mr Gillanders returned to their Saturday meeting with her reply, which heartily congratulated the members of the deputation on their success in London.

> 'Will you also say that I shall be quite pleased, if the Commissioners so desire it, to feu to them the sound so as to enable the Commissioners to construct the necessary work which will give them the loan at $3^1/_2$ instead of 5 per cent. The feu duty would be at the same rate and on the same terms as that for the South Beach. I regret that, owing to severe indisposition, I could not write before today my congratulations to the Committee on the success of their undertaking.'

The Commissioners accepted the terms and the hatchet was buried.

Alexander Macdonald, the Engineer, was instructed to prepare plans of a breakwater without delay. Tracings were to be sent to London as soon as available. A Public Meeting of fishermen and other interested parties was called to get the 'sanction of the ratepaying community' on the proposed changes to the Provisional Order. Aeneas Mackenzie outlined the details of the proposed improvements and the meeting unanimously approved them and urged that there should be no delay in carrying them out.

In order to manage the traffic on the limited quay space, the Pier Commissioners instructed the Harbourmaster to make some new arrangements. He was to make sure that all goods discharged from vessels which had not been uplifted after three hours were removed, failing which he was to have them moved and stored himself at the expense of the consignee. Furthermore, no fishing boat was to be allowed to approach within one hundreds yards of the south end of the Steamer's Wharf if the Harbourmaster considered there was not enough room in the Inner Harbour. To advise the fleet of this, a red flag would fly from the Harbour flagstaff and the fishing vessels would have to anchor clear of the fairway until it was it lowered.

Finally, on 27th June 1892, the Provisional Order received the Royal Assent and the Commissioners were now free to put all their plans into action. On 5th August they held their first duespayers election to bring their number up to ten members. The breakwater plan had to be abandoned due to lack of funding but a start was made on the south beach works at the points opposite Cromwell Street and James Street.

The harbour improvements provided some welcome local employment. Malcolm Smith and John Macdonald, Carters, Laxdale,

Feu Duty

Out of the blue, in fact out of the west, a new headache appeared in 1893. The solicitor for Miss Margaret Robertson Gilbert of Brooklyn, New York, sent a claim for rent in respect of the pier which formed part of her feu on South Beach and which Miss Gilbert alleged they occupied. They denied the claim but because they needed control over the traffic on the pier in the coming season they offered her a rent of £17 10s for the season on condition that they could sublet.

In his letter John Ross said the Commissioners were 'disposed to become tenants of the yard in front of your client's house'. Her solicitor accepted and the Commissioners promptly let it back for the season to John Bissett, a Fraserburgh fishcurer who had occupied it for years, at the rent of £17 10s provided he did not use any part seaward of the roadway. ❑

Fish Mart

14th March 1893 was the day that the Secretary of the Pier Commissiion was able to report that the Public Works Loan Board had authorised a loan of £30,000 repayable over 40 years at 4% interest.

That same day a large number of fishcurers, buyers and salesmen met in the offices of the Fleetwood Fish Carrying Company to discuss a site for a new Fish Market. John Mackenzie Morison explained that the trade had been pressing the Commission for new premises and they were anxious to do all in their power to help. During the ensuing discussion Mr Sutherland, one of the fish buyers, said he thought the South Beach would be the most convenient site. He suggested there should be a place concreted and reserved especially for white fish as the way in which they were presently thrown in heaps was a disgrace to the community. One of the salesmen, Mr Thomson, felt the present stance would be best and Commissioner John Mackenzie pointed out that a plan for another building on this site had been submitted to Lady Matheson the previous year but she had objected to it 'stating that she feared the offices would be let to fishermen and kipper girls'.

As the laughter which greeted this remark died down, he said that the Harbour Commission had long recognised the fact that an auction mart was much required but, owing to pecuniary circumstances they had not been able to remedy the defect.

Mr John Morison, fish salesman, moved that the auction mart be erected on its present site, saying that the South Beach was 'so exposed to south westerly gales that the houses were battened down like a ship's hatch to prevent them being carried away'. The motion was unanimously carried but it would be over a year before the building, which was to become a feature of the Stornoway waterfront for almost eighty years, was finally erected. ❑

gave the best offer for carting dredgings and shingle from Cromwell Street Quay to the new embankment and Alex Barclay, The Dormitory, Stornoway, was employed as timekeeper on the men at the works. Four offers were received for the supply of 300 tons of stone and Malcolm Macdonald, Seaman, Stornoway quoted 1s 5d per ton. The letter of acceptance is addressed to Mr Malcolm Macdonald, (Calum Lochs) Stornoway. At one point, according to a press report of a Commission meeting, 136 men were employed on the foundation work. This number was considered necessary to get as much as possible completed during the very low tides. It was later reduced to 90.

A huge amount of infilling was required on South Beach and the Harbourmaster was instructed to employ as many boats as possible to 'convey filling in stuff from the Inner Harbour and the Sound of Coul Island . . . provided the cost does not exceed one shilling per ton'.

At a Public Meeting held to secure the consent of the Sandwick crofters to a roadway going through their grazings on the north side of the public road running past Sandwick Manse, John Norrie Anderson, set to become a familiar name in the town's history, suggested a solution to two problems. He said that since the Pier and Harbour Commission needed a very large amount of material for their embankment and since the Point community had always wanted the steep incline at Oliver's Brae reduced, here was an opportunity for both. He thought the Commission would be willing to give very reasonable wages for picking, digging and carting the material now on the road.

Stornoway at loggerheads

The *Highland News* details a lively discussion at the November 1893 meeting under the heading 'Stornoway Harbour Board – At Loggerheads Again' during which a Commissioner suggested abandoning the work on the South Beach wall meantime, now that the days were shorter and they would have no banking material to put behind it.

Captain Morison intervened to say he expected to have at least a cart road completed behind the wall before the start of the fishing, just as they had at the slip end that summer. If they stopped the work now they would have neither a wall nor an embankment finished until after the fishing season. They agreed to continue and instructed the Secretary to order the required cement.

A month later Captain Morison was offering to retire if they got a better man to take his place, although he would defy the Commissioners to get the work done more expeditiously. This outburst was prompted by John Mackenzie, shipbroker, asking if there was any report on how much the South Beach work was costing. He said he often went down and watched the men working at night and it was disgraceful how little was done. If it was his money that was being thrown away, he should think it was about time he was being sent to an asylum. Duncan Maciver supported him in this, saying that the men were paid 5d an hour but half of them didn't work a penny worth.

John Mackenzie Morison disagreed. He was sometimes down at the quay at 2 or 3 o'clock in the morning and the men were working as well as during the day.

It was at this point that the Harbourmaster had erupted, saying that he had charge of the men during the day and up till midnight when the Engineer took over. Getting down to the foundations was difficult and sometimes the tide returned and filled the trenches before they could get the cement laid. This was why it was essential to have a large squad of men on hand to fill the trenches during the short time the tide was ebbed.

The shipbroker hastily said that he was not referring to Captain Morrison but to the lack of system in the works, a statement which did not appease either the Harbourmaster or the Engineer. Peter Macleod backed the men and the officials, saying that the state of the tide depended greatly on the wind and that neither Captain Morison nor any other man could tell how far it would ebb. There would be a certain amount of 'scamping' at any works but there was a good job being done. He had seen men waiting to see if they would get a job. Sometimes they would and sometimes they would not. Even if they did get taken on they might only get half an hour's work and he thought the Commissioners should have some sympathy with these men. Duncan Maciver, however, said he had watched the men both at night and day work and he was fairly disgusted with them.

John Mackenzie, fishcurer, felt that they should have put the remainder of the work out to a contractor and said he had lost almost all interest in the matter but Peter Macleod maintained the Commissioners were getting a good substantial wall, which a contractor might have done cheaper but not so well.

The Minute of the meeting simply states that the Engineer was instructed to prepare a report of the exact costs of the South Beach works from their re-commencement in August. The *Highland News*, on the other hand, gave the details of the spat verbatim and ended by saying that a long and exciting discussion afterwards took place on the question of arrears which the paper would print the following week.

The Close Time

Factions were not confined to the Harbour Commissioners. Since the poor fishing of the previouis summer a deep rift had grown between the fishcuring and business community and that of the fishermen. The difference focused on the date on which the summer fishing should begin at Stornoway. On 23rd April 1869 Murdo Macleod, Bank Agent, had written in his diary that seven boats had arrived in the harbour. 'They have come too early,' he said, 'and this opening of the season before the fish is fit for curing will tend to do a great mischief to the market but the poor fishermen are so ill off at home that they are forced to it.' Eventually a 'close time' had been observed in early summer, during which there was no fishing for herring in order to give the fish a chance to mature and to fetch decent prices on the foreign markets. This period was observed by all branches of the trade.

However, in 1893 the fishermen had decided that the period would not end until 15th May and a large body of opinion held that this accounted for the disastrous results that summer. Before the introduction of the close time, the fishing would have started north of the Butt of Lewis in late April or early May and the herring shoals would then have moved down through the Minch, giving Stornoway its livelihood for the rest of the summer. The exceptionally warm weather in the early Spring of 1893 meant that the fish matured much earlier and moved eastwards to the Orkney and Shetland fishing grounds before the Stornoway operation was allowed to begin. It was maintained by some that many in the trade realised that the shoals were moving away much earlier than usual but were bound by the close time agreement not to harvest them until it was too late.

In the Spring of 1894 the fishermen had again agreed that the close time would run from 15th April to 15th May. It was alleged, perhaps unfairly, that the Lewis men needed the extra weeks to prepare their crofting land and that that was one reason why they were so reluctant to start earlier. The curing trade, as well as the business community in general, worried that the delay would mean another bad year. In March the Harbour Commissioners had appointed three of their number – John Mackenzie Morison, Kenneth Smith and John Mackenzie, fishcurer – to approach the Lewis fishermen with a view to coming to an amicable arrangement about the date. The meeting had been held on

30th March but no headway was made. By this time the Burgh Commissioners had become sufficiently alarmed to call a Public Meeting for 4th April in the Drill Hall.

Provost Donald Smith presided over the 'large and enthusiastic' gathering and opened the meeting by saying that, although he believed in a close time, he could not shut his eyes to other interests. When they had set the 15th May date several years earlier, they had the ball at their feet, he said. 'We had no competing ports then. We stood alone as the only port on the West Coast.' Since then other ports to the north of Stornoway had been taking advantage of the close time and were preparing for an early fishing. If Stornoway delayed opening until 15th May, the fishermen would simply go to these other ports and were unlikely to return to the town. They would be driving the trade from their own shores and casting it into the lap of these other ports.

The Provost then called on John Norrie Anderson who spoke in the same vein, again speaking of the benefits of a close time but warning that leaving the start of the season so late could destroy another year's trade at the port. An attempt had been made to mediate between the fishermen and the curers but what had been the fishermen's response? That they had nothing to do with bakers and merchants, that it was a fisherman's question. The Burgh and the Harbour Commissioners had tried again, spending three hours '*in reasoning out the matter*' with the fishermen on Friday but getting nowhere. He said it was time for action:

> 'We are entitled to our opinions and we are going to protect the Lewis fishermen from themselves. . . There is at present a crisis in the history of Stornoway and if no action is taken. . . I would not like to say what will be the result for this town.'

The next speaker was John Mackenzie Morison who suggested that perhaps many of the fishermen were not that much in favour of the close time and were simply following the dictates of a few strong leaders. He described how he had been one of the delegation from the Harbour Commission who had spoken to the fishermen.

> 'I tried to show them that the interests of Stornoway were identical with theirs. Any reasonable man must see that when Stornoway was prosperous the country districts must of necessity be prosperous also. It showed them that if Stornoway comes down it is not to Bayble and Portnaguran the trade will go.'

He thought it was time to assert their position. He would much rather be at war with the Russians than with the men of his own island but he was not prepared, as a native of the town, to stand by and see rivals to Stornoway in the future growing up here and there.

Aeneas Mackenzie, (who, said the *Highland News*, was heartily cheered on mounting the rostrum) accused the Lewis and Harris Fishermen's Association of declaring war on the town of Stornoway and said that if they did not fight now they must sit back and watch their town fade away – 'see the foremost fishing station in the world disappear' – (loud and prolonged applause). 'Are you, gentlemen, prepared to do that?' he asked. 'No, no,' came the response from the body of the hall and Aeneas moved a motion declaring the port open without restriction of landing and inviting all fishermen to prosecute their lawful calling without fear or hindrance. The press report states that much enthusiasm was displayed during the reading of the resolution.

John Mackenzie, shipbroker, seconding the motion made an emotional speech, during which there were bursts of applause. 'It appears to me,' he said, 'that the duty which devolves upon every citizen is to defend his hearth and home.' He continued:

> 'I feel if I don't do my little part it may mean that I will see my children crying for bread and I with no bread to give them. This is not an aggressive meeting; it is a defensive meeting. This matter has been forced upon us after every effort was made to come to a settlement. It is our duty to bind ourselves together – to bind ourselves together in defence of ourselves, our property, and our little ones. If we allow the trade to pass from us with arms folded it will be nothing less than suicide.'

The resolution was passed 'with acclamation' says the *Highland News*, and then John Mackenzie, fishcurer, spoke. He claimed that the fishermen were not unanimous in their feeling on the close time and quoted from a letter from Alexander Thomson of Tong whom he described as one of the best fishermen in Lewis, with a boat and nets second to none in the nation. Mr Thomson was opposed to the delay in opening the port and believed that the commencement date should be 7th May, the same as all other ports. In the course of his speech John Mackenzie also referred to the reasons why the close time had been started years before:

> 'The object was . . . to keep immature herrings out of the Continental markets. At that time we had no kippering houses to speak of, and no freshing trade to boast about. Now we have fourteen kippering houses in full swing and a large freshing trade which together could easily use from 2000 to 3000 crans of herring per day.'

The following week the *Highland News* printed replies by the fishermen claiming that the report on their meeting with the deputations had been one-sided. A letter to the press by three

Sheshader fishermen quoted John Norrie Anderson's remarks about the interest the Harbour Commission had to pay on the loan for improving the harbour and said that 'all the speakers spoke in the same strain, thus conclusively proving that it is for their own interests they are working and not for those of the poor fishermen.'

The letter also quoted Donald Mackenzie, treasurer of the Lewis and Harris Fishermen's Association, who had reminded Mr Anderson of the loans amounting to £12,283 11s 9d advanced to the fishermen to buy boats, which had to be paid back. John Norrie had admitted that this could not be done if the fishermen had to give away their catches for next to nothing and had paid tribute to Mr Mackenzie as 'a long-headed fellow who always worked for the interests of his fellow countrymen in other matters as well as the fishing industry.'

The writers blamed the interpreter for what they said was an 'inaccurate and curtailed' report of the meeting with the deputations from the Harbour and Burgh Commissions. (Presumably the fishermen had presented their case in Gaelic and it had been translated for the benefit of the non-Gaelic speakers at the meeting).

'It is not surprising,' the letter concluded, 'that, like the serpent that closed its ears, he should be provoked and humbled to see a poor, uneducated fisherman more than a match for professed gentlemen and scholars.' The interpreter was John 'Hall' Mackenzie, who eventually settled in Portland, Oregon and whose son, Aeneas, went on to become a Hollywood scriptwriter.

Running through the accounts of this dispute there are several different strands – fisherman versus merchant; working class versus middle class; and sometimes town versus country. Looking beyond the high-flown speeches at the public meeting, it cannot be denied that all the speakers had vested interests in the success of the herring fishing, and the fishermen saw their defection from the support of a close time as purely selfish and prompted by a desire for profit. On the other hand, it was also true that the success and stability of the buyers and curing firms meant employment and money for the skippers and crews, carters and coopers, gutters and packers, from all parts of the island. If the shoals were allowed to drift away, no one made money. The businessmen were unable to contain their impatience with the fishermen, seeing their stance as intransigent, perverse and dangerous enough to destroy all that had been built up over the years.

The Harbour Commissioners were united on this issue. They did not need another season of poor returns and they joined the Burgh Commissioners in sending a deputation to the East Coast ports to assure the fishermen that the port would be open on 7th May and that they would have every protection in the prosecution of their lawful work.

However, also along with the Burgh Commissioners, they publicly and unanimously condemned posters addressed to the 'Fishermen' which appeared overnight in the town from a group calling itself the 'Vigilance Committee of the Inhabitants of Stornoway'. The tone of the poster was belligerent to say the least. It began:

> 'YOUR LEADERS HAVE DECLARED THEIR INTENTION TO RUIN STORNOWAY'

and ended

> '. . . IF YOU DESCEND TO ILLEGAL MEASURES AND ATTEMPT BULLYING YOU WILL FIND THAT YOU ARE USING A WEAPON THAT WILL CUT BOTH WAYS. REMEMBER YOU HAVE TO MEET WITH MEN WHOM YOU HAVE DRIVEN TO THIS AND WILL KEEP THEIR PROMISES. STORNOWAY WILL NOT BE RUINED.'

The Magistrates were worried enough to consult the Sheriff-Principal and *The Scotsman* reported that an application would be made for a fast gun-boat and at least two swift steam launches to arrive at Stornoway by 1st May and to remain for the season. In Edinburgh at a meeting of the Stornoway Fish Carrying Company attended by representatives of a dozen fishcuring firms, it was resolved to have all their employees at Stornoway on 7th May and to be prepared to 'purchase, kipper and cure herrings from and after that date'.

The Commissioners finalised their own preparations for the season by employing nine Deputy Harbour Masters headed by Peter Macpherson at 35s per week. John Morison Senior and Donald Forbes got 30s per week. Four – John Morison Junior, John Maclennan, Murdo Maclean and A J Macaskill (Night Duty) – received 25s. George Mitchell, as Policeman, received 22s and Robert Macleod got 21s for reporting boats. All nine were issued with uniform caps and George Mitchell also received a tunic as befitted his office.

On the evening of Monday 7th May, the night before the port was to open, about ten boats had arrived in the harbour. By this time HMS *Niger* had been anchored in the bay for a week Very early the following morning people began to arrive on the quayside to watch events since, as the *Highland News* declared, the question of 'close time or no close time would be settled by that day's proceedings'.

When the first East Coast boat arrived about 7.30 am with half a cran of herring she was boarded by some Lewis fishermen and there was a brief 'altercation' with the crew. However there was a strong police presence at the harbour and Superintendent Smith took what the paper does not specify but describes as 'prompt measures to prevent any controversy that might lead to serious results'. By this time a second boat had arrived with two crans and by eleven o'clock the first fresh herrings were put up for auction in the new Fish Mart.

A large crowd had gathered for this and shouts went up for someone to make a speech. The auctioneer, Mr Thomson from Buckie, rose to the occasion and his response is worth recording in full:

> 'Well, Gentlemen, I esteem it a great honour to be asked to make a speech on the present occasion however inadequate I may be to do so. When I look up and see the aerial limits of this magnificent building, it makes me think we have taken a very high flight indeed, and it reflects the greatest credit on the Harbour Commissioners of Stornoway that they have provided such a large and comfortable mart for the use of the trade, and that body of gentlemen deserve our warmest thanks for the interest they have taken in the fishing industry. (applause).
>
> 'Taking in the surroundings, we have not only a larger addition to the salesmen, but we have in our midst some of the largest and most experienced buyers in the country and we have likewise one of the finest fleet of boats that the world can produce . . . (applause) To them, gentlemen, we look as the source of all this prosperity. (renewed applause)
>
> 'When I contemplate this scene I feel as if we were transferred to ancient Rome, and as if I were addressing the people from the platform of one of its amphitheatres after some glorious victory. . . (laughter and loud cheers) I consider it one of the greatest honours conferred on me since I first came to Stornoway to be the auctioneer to sell the first fish to break this close time . . . (loud applause) I have no hesitation in asking you to give a hearty vote of thanks to the Stornoway Harbour Commission and those connected with the erection of this fine edifice.' (cheers)

With that the bidding began and the first lot of herring to be sold in 'The Mart' went for 66s per cran to a Stornoway curer. No further opposition was made to the landing of fish and boats continued to arrive, until by Thursday evening there were around 300 in the harbour.

In two short, punishing years the Pier and Harbour Commissioners had accomplished huge changes. They had carried out arguably the most important and far-reaching development of the harbour in its history. These substantial quayworks of the early 1890s laid the foundation for all the developments that followed throughout the next century. To do that the Commissioners had persuaded government to give them a loan equivalent to £2.3 million in today's terms. They had changed the very constitution of the organisation, widening the

The Saga of McCombie's Shed

In late April 1894 Mr McCombie had removed a barricade from the end of his curing station on South Beach Street and had built a wooden shed partly on his station ground and partly beyond it. All this had been done without permission but Mr McCombie explained that he had been forced to move because his former premises had been purchased, the wooden shed was temporary, and he was willing to pay any reasonable rent for the extra ground he was using. He hoped 'he would be pardoned . . . if, owing to the exigencies of the case, his action might be considered unwarranted.'

This put the Commission in a delicate situation because James McCombie was the fishcurer who had been of considerable help to them in securing their recently procured loan. Mr McCombie then requested a five-year lease of the station he occupied and permission to erect a fish shed, cooperage and barrel store.

In a meeting of the Pier commissioners John Norrie Anderson remarked that the Commission should give every assistance to the curers to improve their premises. He believed that every curer should have a roof on their sheds '*to protect their fish from the sun and their women from the weather*'. John Mackenzie Morison thought every member would support Mr Mc Combie's request but he was not sure if they could legally grant permission.

John Norrie Anderson said that he was sure Lady Matheson would prefer to see the stations covered in. After all, in the past she had objected to the bare-footed, and sometimes bare-armed, women being seen along the beach gutting fish. Mr Morison said that this was a very critical period in the town's history and that it would be a great mistake not to encourage the curers in every way. Referring to Lady Matheson's long absence from the island he remarked acidly that the prosperity of Lewis was in her interests too and that '*her aesthetic taste could not be hurt by what she so seldom saw*'. Finally they agreed to write to her Ladyship and await her answer. ❑

representation on the Board and extending the franchise directly to harbour users. They had built a fish market that would be a landmark for years to come.

To quote one of their successors, Sandy Matheson, they were strategists rather than managers. With all their spats and bickerings freely reported in the press, it is easy to dwell on individual characteristics and feuds and to forget that these dozen or so men, acting as a body, had the courage and determination to make Stornoway into one of the foremost fishing ports in Scotland.

Business as Usual

The five years to the turn of the century saw the herring industry in Stornoway at its highest point. In May 1896 the *Highland News* fishing report was congratulating the Stornoway Pier and Harbour Commissioners on the accommodation at the port, saying that they ought to be 'gratified at the appreciative manner in which East Coast fishermen speak of the efforts made for the accommodation of the fleet, many of them stating that there are few, if any, ports in Scotland that present the same facilities for the working of the industry.' The same report, urging provision for a public convenience for the female workers, stated that there were around two thousand women engaged in the trade that summer.

In the issue of 6th June the same year the paper reported on the annual visit of the Channel Fleet and estimated that, with the crews of the fifteen ships in the squadron, the fishing fleet, fishcurers, shopkeepers and ordinary townsfolk, the population of Stornoway town and harbour could be reckoned to be around 25,000 on the Thursday and Friday evenings that week.

The Continental trade was brisk, with steamers departing for St Petersburg, Stettin and Hamburg. 'On an average,' said the newspaper 'there are seldom less than six steamers loading cured herrings at one time.' In fact the Burgh Commissioners were even involved in a correspondence with the Foreign Office regarding what they saw as the lack of proper accommodation for the discharge of herrings at St Petersburg. The Foreign Office informed them that it had passed the matter to Her Majesty's representative at that port to see what could be done.

The building industry was doing exceptionally well – Angus Bain of Bain and Morrison, Fishsalesmen built 'an elegant new villa' on the new Goathill Crescent outside the Burgh boundary. The Lewis Combination Poorhouse was under construction and Provost Smith was erecting a 'spacious villa' on the sloping site opposite the Free Gaelic Manse. This house became known as 'Springfield' and has been home to more than one Provost in its day. Two tenement blocks for 'artisan dwellings' were going up on Church Street. The building contractor William John Mackenzie intended to build himself a large tenement dwelling at the head of Scotland Street and Messrs Tolmie and Gerry were seeking offers for the erection of a block of shops and houses along their fue on Point Street and Castle Street. All this construction work brought cargoes of building materials into the port, adding to the harbour revenue.

The National Telephone Company were considering introducing the telephone to Stornoway. The scheme would only go ahead if there was enough interest. Their Mr Mackenzie from Inverness demonstrated by erecting a temporary wire from the Fish Mart to the Lewis Coffee House and the *Highland News* reported that some of the gentlemen who had been wary in the beginning were 'coming to see the benefits more clearly'.

The Old Order Changing

Lady Matheson died at her London home on 19th March 1896. She was 77 years old and her death came as a shock to the island. The life of the community had been bound up in hers for over half a century – 'a strange and chequered retrospect' as Sheriff-Principal Johnston said in his tribute to her that week. Speaking of the period of the land troubles, he said it was a happy thought that in her declining years the dark cloud had rolled away and that her last visit to her island home was one in which peace and goodwill reigned supreme.

The *Highland News* correspondent remarked on how her vivacity, hospitality and keen interest in public and estate affairs had surprised everyone that autumn. He ended his report:

> 'It may certainly be said that during her ladyship's last stay in the island she loved so well, she made a most favourable impression on all who came in contact with her.'

The Pier and Harbour Commissioners recorded their profound regret at the loss which the town and island had suffered and delegated Kenneth Smith, the longest serving of their number, to represent them at the funeral in Lairg. A party of fourteen local dignitaries as well as ten members of the Local Volunteers left on the steamer at one am on the Thursday morning and were taken by special train from Strome Ferry to Dingwall. On a stormy day, during which the rain fell unceasingly, Mary Jane Matheson was laid to rest beside her husband. As they left the graveside, each of the mourners was presented with an 'In Memoriam' booklet.

The press reported that there were likely to be changes in the management of the estate, with neither Donald Matheson, the new proprietor, nor his son, Major Duncan, expected to take up residence at the Castle. A nephew of Sir James, Donald Matheson's proprietorship

Methodist Recorder

Writing in the *Methodist Recorder* in June 1896, the Rev Arthur Hoyle described the town to his mainland readers. The reverend gentleman had been sent north by the Methodist Church to give pastoral care to what he estimated were about 500 members of the denomination who had come to Stornoway following the fishing:
'I – who was always more afraid of the sea than any beast of the field – am here on an island that is very long and very lonely, but that is just now and for a very little time, swarming with Methodists. . . By the time this letter catches the steamer there will be a thousand fishing boats in the bay, and every boat carries seven men and a boy. To this must be added all the 'gutters' and 'kipper girls' and fish buyers and what not. Thus this little village is crowded for two months with an additional population of 10,000 souls. . . Here there is nothing like herring and nothing but herring from dewy morn to dusky eve, from the grey of Monday to the heat and worry of Saturday night – herring, herring, herring everywhere; and especially obnoxious they are in the nostrils and under the feet. . .' ❑

A Delicate Matter

Mr J N Anderson in accordance with notice moved that urinals be erected on each of the wharves and that a Committee be appointed to consider and report on the whole matter of conveniences for the Female Fishworkers. Mr Kenneth Mackenzie seconded.

It was resolved unanimously to erect the urinals in such places as the meeting should select but it was resolved on the motion of Mr W G Russell seconded by Mr John Mackenzie, Shipbroker, that no committee be appointed to consider and report regarding the conveniences referred to and that the proposal be dropped. ❑

Cruelty

At the end of 1905 the Pier Commissioners considered a letter from the Society for Prevention of Cruelty to Animals asking that cattle pens be erected on the quays to deal with the animals waiting to be shipped out on the mail boat. According to a letter printed in the *Highland News* from one signing himself W.G., on the occasions when the weather was too rough for the mail boat to sail or the vessel was too full, there was nowhere for the cattle to be kept overnight and the buyer had to look for 'some kind farmer' to give them a field for the night.

It was impossible, W.G. said, to keep the cattle together on the quay and they had to be driven into a small enclosure without a gate, known as the Old Market, and taken from there one by one to be loaded on to the boat.

'If he attempts to tie any of the cattle to the rails near the pier,' he added , 'this is objected to by a man in authority in very strong language.' A Committee was directed to visit the pier and to see what needed to be done to meet the SPCA requirements. ❑

was short lived, lasting roughly five years, and there is very little reference to him in the Harbour records.

He had begun his career with the family firm of Jardine Matheson but resigned his post, unable to sanction the opium trade which was its main base. A deeply religious man, Donald Matheson was Honorary Secretary of the Presbyterian Mission in India and Vice President and Treasurer of the Evangelical Alliance. He spent little time on the island and made the estate over to his son, Major Duncan Matheson.

He did not visit the island until July, arriving to a town and harbour bedecked with flags. A crowd of 1500 to 2000 inhabitants had gathered to meet the *Clansman* and a detachment of forty of the Volunteers formed up on the gangway as a guard of honour. Mr and Mrs Matheson were formally welcome to the island by the Provost and a dozen or so VIPs and then came ashore to the cheers of the Volunteers and the crowd.

Books and Book-keepers

In October 1897 another familiar figure took his leave. William Lees, who had been in the Collectorship since the earliest days of the Harbour Commission, handed in his resignation. He was now around 75 years old and, he said, subject to the frequent illnesses and infirmities of old age.

> 'I find that I am unable to discharge my duties with the energy and close attention with which I would wish to do so.'

The Commissioners accepted his resignation with much regret and granted him an annual 'retiring allowance' of £30 in view of his long and faithful service.

'If my life is prolonged for a few years,' Mr Lees wrote to thank them, ' it will be a useful help to me. I also value it very much as a mark of their approval of my conduct while connected with the business of the harbour.'

The position of Harbour Collector was advertised locally and in the *Northern Weekly*. The Commissioners were looking for a man who was 'of strictly sober and temperate habits to devote his whole time to the duties of the office, and to find security to the extent of one thousand pounds for his intromissions.' There were nine replies and the successful applicant was Captain Joseph Couper, a 'Stornoway boy' who was employed by a British shipping company in India. His letter of appointment at an annual salary of £100 went to the Stornoway address of his wife's brother, William J Mackenzie, the building contractor. Another brother was Commissioner John Mackenzie, fishcurer.

This appointment was to give rise to the most damaging public furore in the affairs of the Pier and Harbour Commission to that date. It would eventually end up in the Court of Session. When Captain Couper arrived a few months later, he set about acquainting himself with the book-keeping system in the Harbour Office. This was obviously a fairly complicated and probably very individualistic one, having evolved over thirty years from the modest beginnings of the organisation to the more modern port of 1898. The scale of charges is difficult to interpret even today, frequently being adapted to take account of the changes in the custom of the harbour. Mr Lees was not a trained accountant and the method of recording collections was probably very largely his own, all done with pen and ink and no access to calculators. The volume of work in the fishing seasons and the haste to get cargo vessels back to sea also increased the likelihood of error.

Captain Couper's attempt to get to grips with the harbour dues showed up several weaknesses in the collection system, particularly in the matter of dues on cargoes. There were several main companies, such as Langlands and David MacBrayne, with regular runs in and out of Stornoway. These were recorded in a ledger which continued long into the twentieth century, known as the Steamers Ledger. The other vessels calling at the port were not entered in any ledger, but an invoice was made up for them on the spot. This was handed either to the captain of the ship, who paid it there and then, or to the local agent of the vessel. The only record of the dues on these vessels was a duplicate of the invoice which sometimes recorded the agent's name or initials.

In going through the Arrivals Book and checking off the payments Captain Couper discovered that there had been several vessels over the previous financial year which did not appear to have paid harbour dues. Unfortunately the greater number of these instances was on vessels for which Commissioner Anderson was agent or consignee.

To further muddy the waters, these vessels had not appeared on the list of debtors given to the auditor at the end of the financial year which meant that there was a loss to the Commission. The Assistant Collector, Mr Campbell, questioned on this, explained that when the Treasurer asked for the list, he had been alone in the office and very busy. He had compiled it only from the ledger and had not been asked to give the Auditor any of the duplicate invoices. He maintained that he had always rendered accounts to Mr Anderson.

John Norrie Anderson, who by this time had replaced Donald Smith as Provost of Stornoway, had begun his working life as an apprentice in the legal office of Donald Munro. He had then moved into business life, becoming a ship's agent and an importer of coal, salt and other goods. In the late 1880s he hit on the idea of attracting business by providing coaling facilities from various 'hulks', vessels moored in the harbour as floating re-fuelling points for the many steamers crossing between America and northern Europe.

He was involved in almost every sphere of public life in the town and island, becoming a Burgh Councillor, County Councillor, Harbour Commissioner and member of countless other public bodies. He retained his interest in the legal profession, qualifying as a Notary Public in 1897 and establishing the still extant law firm which eventually became Anderson and Macarthur. He was tireless, indomitable, and forceful, and the shaping of twentieth-century Stornoway owed much to his drive and supreme confidence. When William Gladstone (whose mother had Stornoway connections) died, the *Highland News* reported that the Provost, being unable to attend the late Prime Minister's funeral, had asked Mr Donald Matheson, the proprietor and superior of the island, to go as his representative 'and Mr Matheson, we understand,' said the paper, sounding slightly awed, 'has agreed to do so.'

John M Morison had also had a vessel on the new Collector's list and he having checked his own books and found it owing, paid the dues. John Norrie Anderson said he was in the same position and was making inquiry to see if things were paid. Both Commissioners held that it was not their fault if they had not received invoices from the Collector's office. They had assumed that the dues had been paid by the skippers.

The new Collector's brother-in-law, John Mackenzie, made a statement quoting figures from the Balance Sheets of the previous five years. He remarked on the fact that steamer traffic had increased over that period yet the revenue had dropped. He pointed out that up to the end of June 1897 there had been 55 calls by 'tramp' steamers and only 18 had paid. In the same period of the current year 70 such vessels had called and 70 had paid.

Mr Mackenzie did not blame the last Collector, whom he believed to have been one of the most honest servants that had ever retired from a public board but he knew Mr Lees 'had often complained that he had not the moral support from the Commission which he deserved in the fulfillment of his duties.' Chamberlain Charles Orrock and Provost Anderson both questioned that statement but Peter Macleod agreed. 'The Collector just felt that he had not the Commission behind him. . . Too many of us are too mixed up with the Commission in our business.' The implication was that Mr Lees had found it difficult to be forceful with Board members who had business around the harbour

It was agreed that the books 'for a few years back should be thoroughly overhauled' and the meeting moved on to the next item which was a letter of resignation from Captain Couper, giving three months notice. It had been tendered on the 11th May, two days after their last meeting, and was accepted with regret. The same advertisement for the position of Collector of Dues was sent to the press.

The Dues Committee met a week later to consider a report from Captain Couper which was highly embarrassing for two of the Commissioners. It showed that in the financial year 1895 to 1896, there were 47 vessels in the Arrivals Book for which he could find no record of payment. Of these 34 were consigned to Mr Anderson and 7 to Mr John M Morison, both Harbour Commissioners. The other six were consigned to mainland firms.

Mr Anderson (who was presiding at a Burgh Court and could not attend) had sent a cheque for £33 0s 8d which he had discovered had not been paid. He was investigating the remainder. Mr Morison arrived late at the meeting and took umbrage at his name being included on the list. 'That's what arises through a wrong system of book-keeping,' he said. Peter Macleod replied, 'Apparently it is not the system of book-keeping but the want of system that makes the mischief.'

The Commissioners were now sufficiently alarmed to instruct the Secretary to find a mainland actuary to go over the books and draw up a new set for the incoming Collector. Mr Aikman, a chartered accountant in Glasgow, wrote to say he was willing to audit the Collector's accounts for a fee of thirty guineas.

In the meantime Charles Orrock and John Ross were delegated to visit Captain Couper and see if they could persuade him to stay on. He said he would for a salary of £200 increasing to £250 per annum. After deliberating this, the Commissioners declined the offer and went back to their original plan. Aeneas Mackenzie had sent a letter hoping the question would not deter the Commissioners from coming to a settlement with Captain Couper. John Norrie Anderson (with a 'hear, hear' from Peter Macleod) said they could discharge their duties without any advice or interference from Mr Mackenzie. John Mackenzie added a chilling little footnote by thanking the Commission for having appointed Captain Couper. His friends looked on it as a providence, he said, since the ship the captain had sailed on for so many years, the *Indian Coast,* and which he had left to come home for the Harbour Collectorship, had been lost with the new captain, first officer and ten crewmen.

From the exchanges at these meetings it is clear that the Commissioners were split on the issue of the Collectorship. Captain Couper's brother-in-law was definitely in his corner and other members seemed to have faith in his ability to bring the office into the modern world. Provost Anderson and others seem to have been less keen on his appointment. His place was taken by the Assistant Collector, Alasdair R Campbell who had joined as Collector's Clerk a decade earlier.

Mr Aikman reported to the Dues Committee a week later and, on the strength of that, he was asked to go back a further two years to 1892. This report came before a meeting in October 1898, at which the Secretary reported that Mr Anderson had paid £128 10s 1d of dues which he agreed were outstanding. A balance was still in dispute. The Provost himself complained that rumours had been going round the town to the effect that he had been seen in the Harbour Office 'overhauling' the harbour books. He had been permitted by Mr Aikman, the Auditor, to go through the Arrivals Book for the information he wanted.

The decision of the meeting was that the report was too bulky to be dealt with there and then. The Secretary was to make out statements of the amounts owing by all the debtors who would then be asked to provide receipts to prove that the dues were paid

The meeting was not finished with the Provost that day. The question arose on the correct rate of dues on a Norwegian steamer which had discharged coal into one of his hulks and then left for a mainland port to pick up a part cargo before returning to Stornoway to complete the load. Under the complex rates system Captain Couper had charged the vessel a total of 9d per ton, while Mr Anderson argued it should be $7^1/_2$d. The Provost was not going to fight his case, he said, they could just sue him for it.

Kenneth Smith agreed with Anderson's interpretation and moved acceptance of his payment; but others argued that the rules said the Collector was right. Peter Macleod took up the challenge and moved that Mr Anderson be sued for the full amount. If this steamer was entitled to pay only 7½d per ton then he would be the last man to ask Mr Anderson to pay 9d; but if 9d was the proper amount he would see him in Jericho before he would let him off.

In the Spring of 1899 John Ross's indifferent health broke again and he died at the early age of 40. The new Secretary and Treasurer of the Harbour Commission was Archibald Munro, 18 Bayhead Street, who at fifty had worked his way from postman, to merchant's assistant, to owner of a fish-curing business, and was now also Clerk to the Burgh Commissioners. His salary was set at £50 per annum. John Macdonald, the Interim Secretary, had offered to do the job for £40 but the vote went in favour of Munro, who would serve the Board for over thirty years. He took over at the time when the clash between John Norrie Anderson and Aeneas Mackenzie was at its height.

In the early months of Archibald Munro's appointment he had to deal with the escalating war between these two 'larger than life' characters, opposites in many ways but with the same fire and determination to prove a point and to win. John Norrie Anderson's family roots were in Morayshire but he had been born and raised in Stornoway, where his builder father had come to work on the construction of Lews Castle among other jobs. Aeneas Mackenzie's father had been the noted 'Captain Dany', Daniel Lewis Mackenzie, whose Stornoway pedigree went back several generations and whose family had had considerable influence in the town over the years. Daniel Lewis is listed in *Pigot's Directory* under 'Gentry and Clergy' and he appears in later census returns as a shipowner and house-proprietor. He died when Aeneas was a small boy.

While John Norrie, the elder of the two by a decade, began his working life with Donald Munro in his legal practice, Aeneas started off in the Stornoway branch of the City of Glasgow Bank. John Norrie moved on into coal, insurance, shipbroking and shipping agency. Aeneas leased the Patent Slip from the Matheson Estate and continued the ship building business begun by William Cook. John Norrie studied to become a Notary Public and set up the legal practice which eventually became Anderson and Macarthur. Aeneas was often at law, suing any individual, public body or company that caused him a real or imagined slight. John Norrie was rumoured to have a roving eye. Aeneas was never accused of that. John Norrie was a Liberal, well known enough in Westminster to be considered for a peerage by the government in 1910. Aeneas was an uncompromising Conservative, chairman of the local association.

Both were married, John Norrie with three sons, Aeneas with two daughters. Both were members of the Local Volunteers, in which John Norrie Anderson became a Lieutenant. Both were deeply involved in public life, on the Burgh Council, the Parochial Board, the School Board, the Parish Council and the Harbour Commission.

At the July 1899 meeting of the Commissioners, John Norrie Anderson set about evening the score with Aeneas Mackenzie. He asked that the Secretary and the Collector should bring in a report on the state of the arrears list. He moved also that the Secretary should ascertain whether dues had been paid on the vessel *Columbus* which had come into the harbour some years earlier with a cargo of timber, part of which was for the new wharves and part of which was exported; also whether another vessel, the *Ripple* had paid full dues. It goes without saying who was responsible for both vessels. John Norrie knew very well that there was no record of payment on the *Columbus* or her cargo because he had written to the new Clerk as far back as May and Archibald Munro's reply, dated 3rd June in the Letter Book, had told him that the Collector had carefully searched but had found nothing.

He then moved that 'in order to protect the interests of the Harbour and the Commission' the owner of the *Columbus* be instructed to move the vessel which he felt was becoming a 'dangerous wreck' and might break up.

Aeneas Mackenzie moved that the owner of the *Robert* be instructed to remove that vessel which was sunk off the same point and just as much a danger to shipping as the *Columbus*. When John Mackenzie, as Chairman, repeated the two motions, John M Morison pointed out, to gales of laughter, that it was just one owner giving notice to the other. Peter Macleod, to more laughter, suggested that 'to square it Mr Mackenzie should second Mr Anderson and Mr Anderson should second Mr Mackenzie.'

This pattern of one-upmanship continued throughout the year. Aeneas replied to the charge of non-payment of dues on the *Columbus* cargo by pointing out that the vessel had arrived almost derelict in 1892 after a stormy passage from America, and part her of cargo had been bought by the Commissioners themselves for use in the new piers. The Commissioners agreed not to charge him for that but presented him with a bill for dues on the balance of 400 loads which he had sold to the mainland. John Norrie claimed that 'the owner of the *Ripple*' had not paid the proper wintering dues on the vessel. Archibald Munro said that he and the Collector had gone carefully through the books and could find nothing to show that such dues were payable. John Norrie accepted this and the matter dropped.

The next move came from Aeneas Mackenzie who took up the question of harbour dues payable on the coal hulks which lay at anchor in the harbour. The business of coaling from hulks had been John Norrie Anderson's brainchild. Over the previous ten years, he had brought four or five vessels into the harbour and used them as fuelling points for the Atlantic-Baltic steamer trade.

At a meeting in February 1900, Aeneas made a long and complicated case to the effect that John Norrie's coal hulks had been escaping harbour dues for years due to an unconstitutional decision taken by the Commissioners in 1890 and to the fact that certain charges on steamers' coaling had never been enforced. As he warmed to his theme and backed it up with strings of dates and figures, John Norrie Anderson made a comment decribed in the news report as 'Umph'. Aeneas went on to state that, by not charging harbour dues on the coal discharged or loaded at the hulks, the Commission was in effect subsidising the coal importer, while the fishing industry was heavily taxed.

The battle between the men was waged in other arenas besides the Harbour but it was early 1901 when the last big confrontation between the pair regarding the harbour dues hit the headlines. This happened in the Court of Session when John Norrie Anderson sued Murdoch Maciver, Aeneas Mackenzie's brother-in-law, for £1000 damages over statements made in a pamphlet addressed to the Burgh Commissioners, the Harbour Commissioners and the ratepayers of Stornoway. In November 1898 Maciver had printed a short but highly defamatory leaflet, which he had had distributed widely, accusing Provost Anderson, Bailie John M Morison and the Harbour Master of misappropriating the Harbour funds and implying that were it not for 'that honest mariner, Captain Couper' they would have got away with it. He said that they had 'no title to the confidence of the public and that their expulsion from all positions of trust has become incumbent on the ratepayers.' Its content was so personal and libelous that the *Highland News* had refused to publish it and wisely so, since three months later, Captain Morison won £100 and a public apology in the Court of Session.

John M Morison does not appear to have taken any action and it is not clear why John Norrie Anderson delayed his for almost two years.

The *Highland News,* reporting on the Court of Session hearing, in the Lord Justice Clerk's summing up quoted His Lordship as saying that this was a very curious case and he did not know that he had ever seen one like it. He rapped all the participants in the matter for their 'laxity' in business affairs – John Norrie Anderson, the Collector, the Harbour Commissioners and the Harbour Master – and directed the jury to the central issue which was whether or not Mr Anderson was guilty of dishonesty; not whether he was dilatory or impecunious or putting off paying. The Judge made clear that he thought Murdoch Maciver had been mad to write the letter in the first place and it took the jury twenty minutes to return a unanimous verdict in favour of John Norrie Anderson and to award him £300 damages.

In a blistering editorial, the *Highland News* condemned Murdoch Maciver's folly in concentrating the 'vials of his ignorantly furious wrath' on the innocent parties and hoped that 'all reckless and careless critics of public men will take a lesson' from the outcome of the legal actions.

However, it could be said that the person who was most hurt by the action was the old Collector William Lees. The image of a lax book-keeping system and an inefficient and careless Harbour Office remains in the public record, used as part of the defence of one strong individual against the spleen of the other, who had also used the Collector for his own ends.

By the time the case had come to court, Mr Aikman's improvements had been put in place which kept track of payments by agents but even forty years later, young Collector's Clerks found a less than hearty welcome when they called at businesses in town to ask for dues. The late Donald Murray, Collector's Clerk in 1937, remembered the change of expression on the shop-keeper's face as he bid 'Good morning' to the lady shopper and snarled 'Yes?' at the hapless clerk sent for the thru'pence or sixpence due on a box of gloves or a case of biscuits. William Lees and his clerks would have been no match for John Norrie Anderson in a hurry.

That is not to say that the Provost would have been deliberately dishonest, but that harbour dues would have been far down his list of priorities. Captain Couper's labours had resulted in this situation being checked, but the zeal with which John Norrie Anderson's opponents fell on the issue took the whole affair into the national domain and brought no credit on the Harbour Commission.

In a letter attached to a copy of the notorious pamphlet in the National Library , Kenneth J Mackenzie, the Ross-shire MP, writes to a Mr MacGregor that Sheriff Burns had told him the Pier Dues question had now been 'dexterously metamorphosed into a Free Church question' although the letter does not elaborate on why or by whom.

The assertion seems to have some truth in it judging by the tone of anonymous verses which had appeared on the local scene at the time. One called 'The Harbour Deuce' was written by 'A Seceder' and poked fun at Aeneas Mackenzie, Captain Couper and Murdoch Maciver, while another entitled 'The Harbour Dues' by 'K McPhun' was addressed to 'An Elder of the N. F. C.' and was aimed at the Anderson camp. However no matter what other issues surrounded the row, the root of it might be summed up in the phrase: 'This town ain't big enough for both of us.'

The Thriving Little Town

While Messrs Anderson and Mackenzie had been engaged in their struggle, a man who had probably been more influential in his day than either of them quietly retired from the Commission. Playing tribute to Kenneth 'California', Bailie John M Morison referred to him as the 'Father of the Harbour Trust', remarking that few people realised how much financial loss Mr Smith had suffered through matters affecting their harbour.

Kenneth Smith's name is virtually unknown in the island's history. There is no photograph or portrait that can put a face to the man and his gravestone simply says 'In loving memory of Kenneth Smith'. Yet he made a huge contribution to the development of the island, sinking money he had made in the gold fields of California into fishing stations round the Long Island. In the 1850s his was one of the four major fish-curing firms in the island. He owned a large number of fishing boats, providing the vessels and the gear for the skippers. Above all, it was his financial backing, business ability and refusal to give in which enabled the feuars of Stornoway to take on their proprietor and win the right to regulate Stornoway for the public interest.

At the time of his retiral in the last year of the nineteenth century, the town was gearing up for the new herring fishing season. Kenneth Smith could look back with considerable satisfaction at the progress made.

For the first few years of the twentieth century Stornoway was indeed beginning to thrive. Reporting in 1901 on the cinematographic pictures taken by Mr Mackenzie, Inverness, of the summer fishing season around the Stornoway quays, the *Highland News* referred to the town as 'Herringopolis'. There was still poverty and hardship in the town – enough for John Norrie Anderson to suggest a soup kitchen might be necessary in the winter of 1904 – but around the same time Ada Goodrich-Freer was writing that Stornoway 'smells of fish and reeks of education'.

Just a few weeks after the Royal Visit by King Edward VII, in October 1902, Messrs Bain, Morrison & Co opened their new timber yard on Lewis Street. Their stock of timber included the finest cut kipper box wood which previously had to be imported from the mainland. It was thought that Stornoway probably used more kipper box wood than any other port in the country. In November the steamer *Gerd* discharged 500,000 feet of staves and a deck cargo of timber, for which John M Morison was the agent, and at the end of the Harbour Commission's financial year two weeks later, the total registered tonnage of shipping excluding fishing craft came to 184,960 tons. This was an increase of 10,795 tons on 1901 and 25,478 on 1900.

In 1903 a large barrel factory and stores were being built on Lewis Street for Kenneth Mackenzie, fishcurer. Fish curing and herring gutting sheds were going up for P & J Sutherland, Portsoy and John Mair & Co, Peterhead. Contractors Ross and Mackenzie were also building a kippering establishment for John Woodger and Sons, Great Yarmouth as well as a cooperage and stores for John Mackenzie, fishcurer, and a large carding mill, stores and engineering workshop at the Patent Slip for Aeneas Mackenzie.

Apart from business premises in the town centre, several new houses were going up along Matheson Road and Goathill Road. The GPO was looking for a site for a new Post Office and Messrs Mackenzie and Macleod were engaged in the construction, on Matheson Road, of the Infant Department of the Nicolson Institute.

The largest project of all was the new Town Hall on South Beach, the foundation stone of which was laid in August. In July, Mr William Maclean, Mason Contractor for the build, was given permission to store stones on the quay in front of the building and to open a gateway in the wall on condition that by the 10th of April 1904 all material be removed and the wall rebuilt.

Incandescent lamps improved the lighting in the Fish Mart and new ventilators were also installed there to compensate for the northern door being closed up. A pleasant event was the presentation by Provost Porter of West Anstruther of a 'handsome, double-faced clock, procured by public subscription' for the Fish Market. Angus Macritchie, the Lamplighter, was employed at 10s 6d per annum to undertake the winding of the clock. An application from the Automatic Machine Company to install a two-penny in the slot machine for matches, cigars and sweets was refused. A wash basin and a drinking fountain were fitted by Ross and Mackenzie.

By this time the proprietorship of the island and consequently the Chairmanship of the Commission, had passed to Major Duncan Matheson, son of Donald and grand-nephew of Sir James. Early in 1903 the Commissioners had requested his approval for two of their number to be appointed to a Committee set up by the Scottish Secretary, Lord Balfour, to consider the social, sanitary and industrial condition of the island. Archibald Munro, writing to Major Matheson, pointed out that the improvement of harbours was the very first item mentioned in his Lordship's document, and stressed that the Commissioners were of the opinion that, on a subject of such vital importance to them, they should be represented on the public committee The Major had declined to agree to this, for reasons that are unspecified, and therefore those Commissioners who were already appointed in other capacities were instructed to 'lay before his Lordship the urgent need for a grant towards improving the harbour'.

The early weeks of 1904 saw a very heavy fishing and by February it had broken the record of 36,000 crans landed in the whole of the winter fishing season of 1902. The *Highland News* reported an immense body of fish on the west side of the island which 'for density, is unparalleled in the history of the port at this season.' The Portgordon drifter *Speedwell* came on so sizeable a shoal off the Butt that she could only haul four of her drift of sixty nets. All the others sank to the sea bed. Provost Anderson, seconded by Aeneas Mackenzie, moved that the Harbour Commission support the Town Council's protest against the Close Time for herring fishing from 10th April to 10th May and declare the port open all year round.

The summer fishing was also heavy but the curers were being more cautious about buying. The *Highland News* described the town in their issue of 4th June:

> 'Round the quays from Bayhead to the Slip, in the curing yards throughout the town, and in kipperdom at Newton the place has been a hive of industry. Both as regards daylight and work it may be truly said there is no night in Stornoway at present. The fleet operating here now numbers 660 boats, including 64 steam drifters, about a dozen of the latter having come up from the Donegal coast.'

David MacBrayne's new vessel the *Sheila*, named after the heroine of William Black's novel *A Princess of Thule*, came on the mail run that year. The crowd on the quayside watching her arrival on her maiden voyage raised a cheer when they saw the familiar figure of Captain Colin Cameron on the bridge. Another vessel, the steamship *Trojan* had a more hostile reception when she brought in a cargo of machine-made barrels that month. Unable to find a broker to handle the cargo because of the attitude of the local coopers, Captain Leask tried to discharge some of the barrels himself, watched by a crowd of two or three hundred people. The carters and stevedores hired for the cargo discharge had also refused to work. Captain Leask threw about half a dozen barrels on to the quay only to have them kicked over the side into the harbour. The police were called and thirteen coopers were charged with mobbing and rioting. Sheriff Squair had some sympathy with them and, taking into account their unblemished records and that no violence had taken place, gave them the option of a fine or prison, ranging from £10 or sixty days down to £1 10s or twenty days. The fines were paid.

The saddest arrivals at the quayside in 1904 were the boats bringing in the survivors from the emigrant steamer *Norge* which sank at Rockall on 28th June. Of the 164 men, women and children rescued after six days at sea in open boats, 101 were landed at Stornoway. Reporting on the arrival of the German steamer *Energie* with seventy of the passengers and crew, *The Scotsman* said that a sadder sight could hardly be imagined. Not one of the women and children sent ashore in the first boat could walk up the landing stage without assistance. Shivering and grief-stricken, they were helped to the Fish Mart by local people before being taken to the Lewis Poorhouse.

Stornoway took the *Norge* victims to its heart, providing food, clothes, money and accommodation until they were able to leave. As the mail boat left the pier with the departing emigrants, they lined the rails, waving hats and handkerchiefs, and, said *The Scotsman*, raised rousing cheers which were responded to from the shore.

Some of the survivors of the *Norge* disaster on their way from Stornoway to Liverpool on the *Princess Victoria* 22nd July 1904.

Stornoway Historical Society

John J Maclennan

John Maclennan, 'Iain Alasdair Bhan', buying fish on North Beach Quay in the 1930s; grandfather of John J Maclennan, Harbour General Manager sixty years later

Steam drifters gradually replaced the graceful Zulu-type fishing boats. In 1931 the summer season saw 360 steam drifters, 104 motor boats and only 11 sailing boats fishing out of the port

both photos – Sandy Matheson

James MacGeoch

Norman Mackenzie

James MacGeoch

James MacGeoch

The *Sea Scout* hauling and sorting the catch, August 1949

The photo bottom left on the opposite page shows the *Muirneag* and the *Provider* alongside the quay, date much earlier than the *Sea Scout*

John Macqueen

Barrels laid out on the curing stations on North Beach Quay

Stornoway Fish Selling Co 1924
Front row: W Mackenzie 'Winker'; Albert Thompson; Louis Bain; Ex-provost John Mackenzie; John Mitchell; Fred Woodger.
Back row: William Conder and Mr Brown, (LNER); John Murdo Ross; Mr Butcher (LMS); Bill Keig (Keball); Norrie Maciver

Mrs C Maciver

A Cold Wind Blowing 1914–1939 4

IN 1914 the economy of the town and the island depended almost entirely on fishing. Harris Tweed production was still in its infancy and while the Town Council had been attempting to promote Stornoway as a health and holiday resort, the tourism industry was still very small scale.

Of the ten members of the Commission at the beginning of 1914, the employment of only four was unrelated to fishing – Ossian Robertson was a banker, William A Ross was a solicitor, Donald Maciver owned a shop, and Charles Orrock managed the Lewis estate – although it could be argued that at least some of their daily bread depended on the industry.

Of the other six Commissioners, all were heavily involved in some way in the fish trade. John Mackenzie and Donald Smith both owned long established and well-known fishcuring businesses. Duncan Maciver, who had been a young Commissioner for a brief spell in the 1890s and who was returned to the Board by the dues payers in 1911, had become a highly successful fishcurer and boat owner in the interval, building up what became possibly the most famous of all the Stornoway kippering firms. Bailie Murdo Maclean owned a number of fishing boats and was the first local owner to purchase a steam drifter, the *Lord Methven* in 1910. Kenneth Mackenzie had begun as a fishcurer and, although by 1914 he had become involved in the tweed business, at the beginning of the war he was still a shipbroker, coal merchant and an exporter of fish, with outposts in the Shetlands and on the east coast of England. John Norrie Anderson, that man of many parts, was listed in the Minutes as a Solicitor but was also a steamboat agent, a coal merchant and an importer of, among other things, the large cargoes of salt needed for the fishcuring process.

The start of the new year of 1914 was a promising one for the Stornoway Pier and Harbour Commissioners. Provost John Mackenzie, speaking at their meeting on 2nd February, outlined the success of the winter's fishing. At that point in the season the catch of 61,000 crans was up on the previous year by 18,000 crans and over seventy buyers were working on the quays. Around 10,000 boxes of kippers were being exported per day, not to mention the fresh herring which was being shipped. Three days earlier some of the fish had had to be sent to other fishing stations because there was not enough room on the Stornoway quays to deal with all the catch in one day. The Provost was concerned that the quay space at the port was once more becoming inadequate for the trade.

Therefore when Provost Mackenzie moved that the Commission revive the plans they had had in 1892 to build a quay and breakwater between Goat Island and Newton, his fellow Commissioners were all in agreement. Financially they felt that they were in a good position. They had paid back £28,000 of their £45,000 loan over the intervening years and it looked as though their income in the current financial year would be more than the previous year's £8000. On one day in the winter season, according to the Provost, the harbour had held 250 steam drifters, 3 motor boats and 18 sailing boats and, as John Norrie Anderson, seconding the motion, pointed out, the port's accommodation was inadequate for that level of traffic.

Mr Anderson also reminded the meeting that Stornoway Pier and Harbour Commission had never received a Treasury grant for harbour extension, as some other Scottish harbours had and therefore perhaps they should apply to the Development Commissioners for a grant or a 'free loan' for the extension work He had also suggested that they should do something about deepening the Inner Harbour, the funding of which he thought the Commission could manage themselves. The matter was referred to a Works Committee consisting of the whole Commission to look into what needed to be done, and the recently retired Captain Morrison, along with the new Harbour Master Malcolm Maclean were co-opted to assist in the deliberations. Provost Mackenzie tossed in one idea to be going on with, that of constructing a drawbridge over the Patent Slip to allow free access to Newton.

From weighty matters of new piers and Government loans, the meeting then moved on to consider the problems of finding assistants for the Harbour Master, the poor lighting on No 2 and No 3 piers, and the Weighmaster's complaint about the 'quite unsuitable' oil lamps in the weighbridge. These were dealt with by giving Provost Mackenzie authority to hire experienced men at the East Coast fishing ports, deciding on the use of incandescent lamps at the said piers, and instructing that a gas pipe be carried in to the weighbridge, a 'suitable' bracket fixed, and a penny in the slot meter installed. The Weighmaster was to provide the pennies, being required to pay for the gas consumed.

In fairness to the Commission, the Weighmaster, Captain Kenneth Maclean, also received some of the takings of the weighbridge under the peculiar arrangements in place in 1914.

The Works Committee set about their task immediately. Archibald Munro produced the plans of the extension abandoned in 1892 and Alexander Macdonald was engaged to produce measurements and costings along with the Harbour Master and Captain Morrison, including a special report and estimate for extending the Inner Harbour quay 'from the north end of the Commission's property on Cromwell Street to the steps at Bayhead'. The Secretary was to advertise for a self-propelling bucket for the dredger with a carrying capacity of 200 to 300 tons.

In the meantime the first rumblings were being heard of a long dispute with Messrs MacBraynes. The new Collector of Dues, Angus Macleod, who had replaced AR Campbell in 1912, had informed the Commissioners in January that he had been charging dues on that company's vessel *Chieftain* at 306 tons which was the figure 'as returned to the Harbour Office'. However the vessel's official tonnage, which he had now established, was 604. Angus Macleod asked for instructions as to what to do about the balance of £803 17s 6d outstanding since her first arrival on 8th May 1898. The Commissioners decided to 'make a call for payment' on Messrs MacBrayne and instructed the Collector to make out an invoice.

Archibald Munro sent the invoice to David Simpson, the local agent for David MacBrayne, and, having explained the situation, said the Commissioners would be obliged by an early settlement. Mr Simpson sent the invoice back. Mr Munro inquired if Mr Simpson had sent it on to his principals, otherwise he would have to send it to Glasgow himself. Four days later Mr Munro sent the letter on to Glasgow. Three weeks later he wrote to Hope Street again saying he would be obliged by having a definite reply 'as I assume you have gone into the matter by this time'. On 9th March he was writing to say he was astonished that he had received no communication from them. On the 16th March he was writing to David Simpson again, agreeing that the Harbour claim required some time to look into it but 'two months should have more than sufficed for this'. He sympathised that Mr Macbrayne had been ill and hoped he was now at business and able to give the matter his immediate attention.

David Simpson had written directly to the Collector to complain and Archibald Munro stressed that

> '. . . he has not called on you to pay an increase of dues, all he has

> done is to charge the same rates of dues as formerly charged but on the correct tonnage, the accuracy of which is beyond doubt and there seems to be no good reason for withholding payment of the last three months accounts, although the other claim is under your consideration.'

MacBrayne's withheld their payments for many months and the tussle went on for almost the duration of the war which, unknown to them, was just about to break out.

Into the Cloud

Less than two weeks after the meeting of the Commissioners in June 1914 the dark cloud of war began to roll across Europe, beginning in Sarajevo with the assassination of Archduke Francis Ferdinand and the resulting conflict in the Balkans.

In the *Highland News* issue of 1st August 1914, John Norrie Anderson, as Vice-Consul for France, reported that the 3rd Class Cruiser, *D'Estres*, with its complement of naval cadets, would visit Stornoway on first week of August but on the same day as his words were printed Germany was declaring war on Russia and would follow this with a declaration of war on France two days later. The newspaper was reporting elsewhere in its columns that 'the outbreak of hostilities has completely paralysed the fishing industry.' Every night hundreds of fishergirls were coming back to the island from the East Coast. One night that week the *Sheila* had carried 200 passengers and MacBrayne's had sent their steamers *Gael* and *Chieftain* to assist. Between them, these two vessels brought over a thousand Lewisfolk home to the island that night.

As the women came back, hundreds of their men folk were leaving, called to service from the Reserve Forces to which so many island men belonged. Even before Britain's declaration of war on Germany on 4th August, the island was almost emptied of its able-bodied men. On Thursday 30th July the Stornoway Company of the Ross Mountain Battery had left Stornoway for Dingwall under the command of Captain Thomas Nicolson. The *Highland News* recorded that they were given a magnificent send off.

In these very early days of the war it was felt that Stornoway was not likely to be affected by the hostilities and therefore some of the boats which could not fish the North Sea came round to the Minch, while the kippering premises reopened and began their business again. Archibald Munro's letters begin to reappear in September, dealing with a whole new set of everyday problems.

A new element in harbour traffic was the detention of various foreign vessels on their way across the Atlantic. These were detained by the naval and customs vessels stationed at the port and having been searched and cleared, were allowed to continue on their way. Any crew or passengers belonging to the enemy nations were taken ashore to the Custom House and then despatched south to internment camps. One such vessel was the Danish liner *Oscar II* on passage from New York to Christiana and Copenhagen at the end of October 1914. She had several German diplomats on board who were allowed to continue, although six German crew members were taken to the Drill Hall until they could be shipped away. Also on board that night but undetected was the Irishman, Sir Roger Casement, later executed as a traitor.

In mid December a notice appeared on the Fish Market door declaring that fishing was to be restricted to fishing boats registered in Stornoway holding permits from the Fishery Board counter-signed by the senior naval officer of the division. This meant that the East Coast boats could not fish in the Minch and consequently the winter fishing season would collapse. Provost John Mackenzie and John Macritchie Morrison, fishsalesman, left for the mainland, going first to Aultbea to speak to the Rear-Admiral of the Fleet and then on to the Fishery Board in Edinburgh but to no avail. The herring fishing stopped and with it the income of its hundreds of women workers.

One piece of correspondence at the February 1915 meeting was a 'confidential communication' from Rear-Admiral Tupper. By this time the port had become a naval base, with the name of HMS *Iolaire*, and Rear-Admiral Tupper had been transferred from Aultbea to take charge of the military operations. His letter outlined in detail the Navy's plans for safeguarding the anchorages of the port and now requested the Commissioners to submit a scheme for 'rendering the port wharves etc useless to the enemy'. The Rear Admiral added that the co-operation of all connected with the port was desired in order to ensure the effectiveness of the orders.

The Commissioners, after some discussion, decided that the most effective means at their disposal would be to 'place three or four barrels of petroleum near to each of the three wooden wharves in readiness to saturate them therewith for the purpose of firing them if an occasion should arise to render them useless to the enemy' and Archibald Munro enquired if this scheme met with the Rear-Admiral's approval. It did, according to the Minute.

Customs activity at the port was stepped up because of the war and their office accommodation in the Fish Mart was increased to house the extra members of staff. All the coal hulks were moved to Glumaig Bay to allow space for the steamers being detained for investigation. Some were allowed to continue on their way, others like the *Forsvik*, bound from Vigo in Spain to Gothenburg in Sweden with rice, were placed in what was known as the 'Prize Court' and sent to the mainland for detention. Towards the end of the year one of these requisitioned vessels came into the port with a shifting cargo of timber. She was now the *Parklands* but her previous name had been the *Marie Gläser* and she was the first interned German ship sold after the outbreak of war.

Throughout 1915 and 1916 many rescued sailors of different nationalities passed through the port as a result of U-boat activity on the west coast of Lewis. In one week in July 1915 eighty men from five sunken vessels were sent home. In October that year Stornoway Harbour was closed at night from one hour after sunset to one hour before sunrise. All vessels, apart from local ones, arriving at the port within these hours had to go to anchor until the appointed time and then report to the guard vessel that sat outside the harbour before going any further.

At the Annual General Meeting Kenneth Mackenzie had made a motion to have the necessary repairs to the harbour carried out. He said it was time to seriously consider the dredging of the Inner Harbour since the channel opposite the corner of Esplanade Quay had silted up so much that there was now a bar across it, preventing boats getting to the inner quays even at half tide. He suggested that the unsightly wrecks lying at Bayhead should be towed out to Goat Island and when peace was declared, they should celebrate the event by having a bonfire of them, assisted by the barrels of tar at present lying near the quays. Mr Mackenzie stated that the harbour was the best asset the town had (a view he frequently expressed in public) and the least they could do was try to keep it in good repair. The Commissioners agreed with him and instructed that Alexander Macdonald be engaged to inspect the wharves to see what work needed to be done.

Wartime or not, the dispute with MacBrayne's had lurched on through its second year. At the beginning of 1915 the Commissioners had decided to rescind the decisions taken by previous Commissions over the years, giving compounded dues to Messrs Langlands and to charge both Langlands and MacBrayne's on the rate set in 1895 for steamers arriving from home (i.e. British) ports, discharging and then loading and sailing for home ports. They had taken legal advice on the situation and in May the Harbour Collector Angus Macleod had been set the task of calculating the difference between what MacBrayne's had previously paid on certain ships and this compounded rate. This resulted in the debt on the *Chieftain* being reduced from £803 to £368 and the Commissioners hoped that this longstanding dispute would now be settled.

MacBrayne's enquired as to how the sum was arrived at and the Commissioners, feeling that the Clerk had already explained quite clearly, demanded settlement within a month or legal steps would be taken. By September, no reply had been received and the fact that MacBrayne's had not paid any dues at all for two years was beginning to be discussed among the other traders at the port. Kenneth Mackenzie, speaking as a duespayers representative, said it was time to force the issue. Provost Mackenzie said they should separate the two issues and insist on the payment of the account for current dues, which was not in dispute, even if they had to raise a court action.

The Clerk having been instructed to wire for a reply by first post, MacBrayne's wrote back to say they had never received his July letter which was not their fault and requested that 'copies of the figures arrived at be sent to them so that they may check them against their own account and further inquiring if the Commissioners are now complying with the terms of their Act.'

The Commissioners sent them the figures but asked for settlement within 14 days. Three weeks passed before MacBrayne's replied that their solicitor had all the papers but was away from home for a few days and they would be in touch when he returned. The Commissioners deferred the matter until their monthly meeting in November, by which time they had heard nothing. On the motion of Kenneth Mackenzie seconded by Donald Maciver, they decided to raise an action for the recovery of the money in the Court of Session in Edinburgh.

By their December meeting they had received a letter from MacBrayne's in which the company stated that, while they were advised that their stand was legally sound, they wished to maintain a good relationship with the Pier and Harbour Commission, and suggested that a deputation of Commissioners should meet them in Glasgow or Inverness to 'frankly state to each other what the respective positions are'. The Clerk was instructed to reply that the Commissioners had waited so long for a reply that they had placed the matter in the hands of their solicitors. His letter also informed MacBrayne's that, in view of the fact that they had not paid any dues on any of their vessels for the last two years, the Harbour Collector had been instructed to demand payment from them weekly.

No End in Sight

The turn of the year 1915-16 was cold and miserable. Over these weeks, on the other side of the continent, British troops had been evacuated from Gallipoli and the newspapers continued to list the names of the dead and the missing. The island was earning its name of 'Loyal Lewis', with over four thousand men serving in the forces. *The Scotsman* noted a few months later that Lewis would hardly be touched by the compulsory call-up which was being introduced, since there were very few men fit for military service left at home.

In the Fish Mart, a free Tea and Rest Room for soldiers and sailors had been set up and by this time over two thousand servicemen had passed through its door. The girls who had earned their living gutting and packing the 'silver darlings' on the Stornoway quays were being recruited for the munitions factories in the south, and in February some of them left for Dundee where they were employed to make sandbags. This was the year of the battles of Verdun and the Somme, the Battle of Jutland at sea, and the Easter uprising in Ireland.

Leafing through the Minute Book of the Stornoway Pier and Harbour Commission it is easy to forget the background against which all its deliberations took place and the fears the Commissioners must have had for their sons and for the future. Duncan Maciver and John Norrie Anderson each had two sons in the army. Charles Orrock's son, Colin, served with the Royal Field Artillery and Captain Alfred Robertson, son of Ossian, with the Royal Army Medical Corps. Kenneth Mackenzie's son, Alastair, was a Lieutenant in the RNVR.

There were two main recurring items of business all through 1916. Apart from the escalating dispute with MacBrayne's, the other was the difficulties arising from the Admiralty presence in the harbour.

The Finance Committee recommended that the Commissioners should frame their claim for compensation to the Admiralty for the use of No 1 Pier along the same lines as Lerwick had. In November it was agreed to set up a Committee along with members of the Town Council and the Lewis District Committee to try to get some relaxation in the restrictions on fishing boats entering the harbour. Archibald Munro wrote to the Rear Admiral to ask that he free the east side of No 1 Pier for the landing and shipping purposes of the large number of boats expected for the winter fishing season but the regretful reply was that the accommodation and wharfage for HM ships were already sufficiently limited as to preclude the surrender of any for the fishing trade.

At the same time as fencing with the Admiralty, Angus Macleod had reported at the beginning of the year that, although he had been rendering weekly accounts to MacBrayne's local agent, Mr Simpson, these were being returned to him with the comment that they should be sent to Messrs Moncrieff, Warren, Paterson & Co in Glasgow. Ex-Provost Anderson had moved that the company be asked to settle the £130 17s 5d now owing on the weekly dues by midday on Saturday 12th February, failing which the Commission would invoke their powers to prevent MacBrayne's vessels from berthing or landing or shipping cargo at any of the piers. This letter was sent registered post and a copy supplied to Mr Simpson.

MacBrayne's responded to this move by taking out an interim interdict against the threatened proceedings and the Commissioners set up a Committee to frame answers to the notice served on them. The Clerk, writing to their Edinburgh solicitor, Alexander Ross, admitted that on checking the Act, they had discovered they had no powers to prevent landing etc, although under Clause 44 of the Harbour, Docks and Piers Act 1847, they did have powers to distrain. One of the first actions the Committee took was to have the harbour rates relevant to steamers printed and handed out to all persons in charge of vessels in the harbour. Three or four of these were to be displayed at points on the quays and one outside the Harbour Office. This was all in response to MacBrayne's contention that, since they had not published their Schedule of Dues as required under the General Harbours Act, the Commission had no right to record any dues.

In early March David Simpson advised the Collector that, if he was issued with separate invoices for each of the company's steamers, he would arrange to pay them himself either weekly or after each visit, whichever the Collector preferred. On 14th March the Clerk called a special meeting to consider MacBrayne's withdrawal of their interdict and their renewed suggestion of a meeting between the parties. Mr Ross strongly advised they should agree to a meeting and certainly should not serve any summons on the company until it had been held.

A deputation consisting of Charles Orrock, Provost John Mackenzie, Ex-Provost Anderson, Bailie Murdo Maclean and Mr Roderick Smith was nominated to go to Edinburgh on 6th April to meet first with their legal advisers and then with MacBraynes. A telegram from the company regretted they were not free on the 6th but they could manage the 7th in Glasgow. Archibald Munro did not think his Commissioners would agree to this and a copy of his telegram shows the next move in this game of chess.

> 'DEPUTATION AGREE TO FRIDAY THE SEVENTH TO MEET MACBRAYNE AT EDINBURGH. DON'T AGREE TO MEETING IN GLASGOW. WIRE IF THIS ACCEPTABLE.'

MacBrayne's held out and the next wire to Mr Ross agreed to Friday 7th April in Glasgow.

The meeting was not a success. The gist of the MacBrayne argument was that they wanted all of their old accounts to be backdated prior to 1908 to put them on exactly the same footing as Langlands and they were willing to go to arbitration and enter into a Reference to an Accountant. Both of these proposals were rejected by the deputation and Mr Ross said that the Commission would prefer to have the matter decided by the Court. Mr Warren acting for MacBrayne's replied that, if the Commissioners would agree to a Reference, his clients would drop their contention on the non-publication of the Schedule of Dues. Mr Ross felt this was a very material concession. In a further letter he passed on the opinion of counsel on whether or not MacBrayne's could have the old accounts reopened. Mr Wilton, the Counsel engaged by the Commission, was of the view that, if it could be shown that MacBrayne's were not aware of any preferential treatment that might have been given to Langlands or any other company during that period, they would not be barred from claiming compensation. On the other hand, if they had known of any such preferences and had still paid their dues, they would probably be barred from revisiting the subject.

This meant more work for the Collector who was now instructed to prepare a statement from 1884 to 1915 showing the concession made to Langlands and to MacBrayne's in that period. Having produced the required information at a meeting of the Finance Committee on 22nd May, Angus Macleod was instructed to go back two years further to 1882, and to check and verify all figures in the statement. The Clerk was instructed to check the Minutes authorising the 50% reduction in the mail boat dues, as well the reference to commuting the mail boat dues to £1 per trip. The Committee recommended that the Commission agree to arbitration on condition that all Messrs MacBrayne's steamers were included in the consideration and also that they drop their objection to the non-exhibition of the dues.

So ended 1916 for the Commissioners, a year in which they seemed to have achieved nothing. The start of 1917 was not auspicious either. In mid-January the bad weather damaged the telegraph cable to the mainland, causing disruption to the fishing community and Rear-Admiral Boyle arranged for the patrol boats to take telegrams to and from Kyle twice daily in order to keep the town in touch with the outside world.

For the very first time, the Stornoway kipper girls went on strike for more wages, adopting 'suffragette tactics' observed the *Highland News* sourly, referring to their march through town, singing, and their breaking of a few windows at the kippering sheds. They returned to work after a few days on the same rates as before and the paper implied that this was their own fault, hinting that the whole episode had been engineered by the 'Peterhead girls'.

On the continent the war dragged on through the battles of Arras and Passchendaele, while at sea Germany stepped up her submarine campaign. For a time in 1917, food supplies in Britain ran dangerously low as a result of the success of U-boat activity in the Atlantic and rationing of staple goods was brought in as one of the measures to combat this.

At their January meeting the Commissioners had before them the Opinion of Counsel on the MacBrayne dispute. In brief Mr Wilton and Mr Horne advised that, since MacBrayne's would only go to arbitration if their mail steamers were removed from the comparison on concessions and since it was obvious that the Harbour Commission could not agree to that, it might be necessary to begin legal action. However, this was a matter 'attended with considerable difficulty' and Counsel therefore advised that the Commissioners offer to drop the claim for the backdated *Chieftain* dues subject to MacBrayne's immediately paying all the undisputed dues they had withheld.

Meanwhile MacBrayne's refused to pay the new water charges which the Commission had set at the end of 1916. At the end of January David Simpson wrote to ask for increased storage accommodation, claiming that the mail boat had been prevented on several occasions from discharging all her cargo due to the lack of storage space and suggesting that the Stornoway Shipping Company store be made over to them in the circumstances. The Commissioners held a Special Meeting to deal with the request and Charles Orrock voiced his worry that Stornoway might be starved, presumably afraid that the wrangling between the parties would lead to more serious consequences.

Baillie Roderick Smith replied firmly: 'The Government will not allow the inhabitants of the Island of Lewis to be starved, whatever MacBrayne's or any other carrier may do.'

Two of the members present at the meeting were directors of the Stornoway Shipping Company and they informed the meeting that they had offered to sub-let part of their store to MacBrayne's subject to the agreement of the whole Commission. Archibald Munro wrote to Hope Street to say that the Commission had no storage to offer but would make no objection to any arrangement of sub-let between MacBrayne's and the Stornoway Shipping Company.

This was the period in which there was most fear about the nation's food supplies and in a note to Rear-Admiral Boyle dated 21st April the Clerk acknowledges the information that one of HM drifters is going to bring a quantity of seed potatoes free of cost through Mr J N Anderson, for issue to crofters and others engaged in agriculture. He told the Rear-Admiral that he would bring the question of the waiving of harbour dues on it to the next meeting of the Commissioners but there is no mention of it in the May meeting which was the last to be held until early August that year. That Spring questions were raised in the House of Commons regarding the amount of 'herrings pickled in the brine of Stornoway' which was stated to be 75,000 barrels. Their export was prohibited, since it was considered they should be kept for consumption under the current circumstances.

The barrels were still there on 10th May when the Clerk apologised to Captain Knowling at the naval base for the accumulation of refuse on the pier next to the Chief Boatswain's store. This was because the piers were so congested with herring that there was nowhere else to put the harbour sweepings until a cart could be got.

Light on the Horizon

In early 1918 the outlook on the war front was still bleak. The Germans were able to launch a renewed offensive on the Western Front, having made peace with the new Bolshevist regime in Russia. At that point the island was experiencing shortages of staple goods and there were severe losses through delays in transit of consignments of herring going south. In February 1918 representatives of the Fish Trade met in the Library Room in Stornoway to form an Association 'for the protection of their interests'. In order to free up the traffic, the meeting was inclined to ask the Ministry of Food to keep Oban clear for Stornoway shippers; and, for their part, the traders to stop sending consignments to Mallaig.

Two major events would have taken up the time and energy of the Chairman and the Secretary in February and March 1918. The first would have been the sale of the island by Colonel Matheson, rumours of which had been circulating for some time. *The Scotsman* of 15th February announced the purchase of the island, which it described as the largest territorial property in the United Kingdom, to Lord Leverhulme, owner of the vast Lever Brothers empire. This was not the first time the Colonel had attempted to sell the property, having offered it to Lloyd George as far back as 1913, according to *The Scotsman* report. The successful sale removed an uncertainty for Charles Orrock and the other Commissioners. The future began to look a little bit brighter. Looker-

Random Notes

In November 1917 a new column had appeared in the *Highland News* entitled 'Random Notes' and attributed to one 'Looker-On'. It was a satirical, sometimes biting, comment on events and personages in the town and there was great speculation as to who the author might be. At one point the Editor was forced to deny that it was the Science Master of the Nicolson Institute after comments made at a meeting of the School Board, which presumably would have caused that gentleman considerable trouble. However, the *Highland News* representative in Stornoway happened to be Catherine M Macleod, sister of the Collector of Dues, who had taken over in January 1917 when William Grant had left to found the *Stornoway Gazette* and Stornoway folklore has attributed authorship to her.

The writer reserved their particular scorn for Stornoway Pier and Harbour Commission and it is tempting to think that perhaps Catherine Macleod's brother had a hand in the composition of the column, if he were not himself the culprit after a fierce clash with his employers six months earlier. In the issue of 6th December 1917, having lambasted the Parish Council as 'too rotten for words', the School Board as 'moribund' and the Town Council as 'asleep like a hibernating frog in a stagnant pool', Looker-On continued 'Worst of all by a long way is the Pier and Harbour Commission. It is diseased, it is comatose and no stimulant can rouse it to life again – not even the prospect of losing two or three thousand pounds sterling – as they are likely to do owing to their crass stupidity and indifference in their dispute with our worthy benefactor and philanthropist, Mr MacBrayne'.

The article went on to question whether it was true that the Commission had a large sum of money on deposit in the bank, and if so why it had not been spent on repairs to the quays. Efficiency had vanished with the retiral of the old Harbourmaster and the ratepayers' representatives were doing nothing. ❑

On, in the Random Notes column of the *Highland News* hoped that the arrival of the 'captain of industry' would stir the Harbour Commission into life again.

The second, very different, event came two weeks later at 5 pm on Saturday 2nd March 1918, when fire broke out in the Town Hall. It took a steady hold and by ten o'clock that night it had completely destroyed the beautiful Municipal Buildings so proudly opened thirteen years earlier and 'considered to be the finest of its kind north of Aberdeen'. In its ashes were most of the contents of the Public Library and the furniture and effects of several public bodies, including the Town Council, the Pier and Harbour Commission.

Archibald Munro and Bailie Peter Macleod did their best to rescue the public records and to them the Port Authority owes the salvage of the Stornoway Pier & Harbour Commission Minute Books, Letter Books and Cash Books from which this history is gleaned. The books were charred and water damaged on the outside but largely intact and, through the good offices of the Scottish Record Office, were restored, rebound and microfilmed in 1998. The *Highland News* reported the comment from the Clerk that he might have got more records out but by the time he got downstairs the flames were spreading and it was only with difficulty he was able to get out himself. The Town Council Minute Books had also been saved and all the papers in the Strong Room were untouched by the fire.

The Commissioners joined the Town Council in organising a public welcome for the new proprietor. This took place late on a July evening, so late that it was decided to dispense with the pipe band music which had been planned. Lord Leverhulme and his party were met by Provost Murdo Maclean, Charles Orrock and ex-Provost Anderson, on whose call three rousing cheers went up from the large crowd on the pier. His Lordship passed through a 'triumphal arch' of evergreens bearing the legend 'Welcome to your Island Home' and his short-lived tenure of Lewis began.

In his reply to the official speeches at the Masonic Hall Lord Leverhulme spoke warmly of the Matheson proprietorship and his own wish to be of service to the community, which he said was the only justification for having possession. He painted a picture of the future development of the island that must have lifted the hearts of his listeners, speaking of the wealth in the seas around the island and the potential which modern methods could realise. The Provost, said his Lordship, had referred to Lewis as being considered an outlying part of the world, but in fact, as far as the sea was concerned, it was the hub of the universe. He intended to give the people of Lewis the opportunity to work out their own destiny. *The Scotsman* recorded his closing remarks:

> We shall work on business lines. We shall have nothing to do with philanthropy. We will be able to look each other in the face and you will be able to say to me 'We thank you for the opportunity and that is all we want'; and I shall say to you 'We are good friends. It is a pleasure to live amongst you and I hope we shall have a long and happy life together.'

Lord Leverhulme's lively, energetic approach was a marked contrast to the more restrained manner of his predecessor. Both men had a genuine love for the island, and although Matheson may have had a better understanding of the culture and the people, he had neither the money nor the political will to change the status quo. Leverhulme, on the other hand, had both the money and an iron will to change anything which he thought could be done better.

His first meeting as Chairman looked set to fulfil the hope expressed by Looker-On some weeks before that his Lordship would 'bring his business talents to bear upon the transactions of this important body'.

Lord Leverhulme

The most radical decision at the 23rd of September Commissioners' meeting was taken on the motion of Kenneth Mackenzie seconded by Provost Maclean. It asked that Lord Leverhulme take all necessary steps to procure a Provisional Order on his own behalf as proprietor of the foreshores . This decision prompted alarm in the Town Council where Peter Macleod took the Council representatives to task for their action. Looker-On in the *Highland News* drew attention to the fact that the Press had not been invited to the meeting at which the Commissioner had agreed to support Lord Leverhulme 'in promoting a Provisional Order which, under certain circumstances might result in Lord Leverhulme or his heirs obtaining a pre-dominant interest in the harbour'.

The writer concluded by saying that if the interests of the town were to be conserved the present Harbour Commission must be reconstituted:

> 'It consists of ten members, six of whom are returned by the ratepayers and they should see that no time-serving lickspittles are allowed to barter away their heritage merely to curry favour with the new laird.'

The unstoppable new laird pressed on. At the end of that week he was in Aberdeen, being shown over the Fish Market and the other facilities at that port in view of his intention, according to *The Scotsman*, to make Stornoway 'the Hull of the North'. and on 4th November the

Commissioners heard that he had had an interview with the Secretary for Scotland, Robert Munro, who wished to see the plans of the proposed extension to the quays as well as those for his proposed Railway Scheme. Lord Leverhulme had commissioned Mr Wall, Lever Brothers' Civil Engineer to prepare drawings to place before the Scottish Secretary as soon as possible.

On 11th November 1918 at 11 o'clock forenoon, the guns were falling silent all over Europe.

By the end of the year Lord Leverhulme's 'perfected plans' for the proposed extension and the line of the intended railways had been approved. The cost was estimated at £250,000, a staggering sum for its time, but one for which the Commissioners felt the Government was morally bound to provide funds at a nominal rate of interest. Kenneth Mackenzie had moved the approval of the plans with a long speech in which he called attention to the great sacrifice of the island's young men and the indebtedness of the Government towards Lewis in the late conflict. He believed that, with the increased income which would result from the improvements, the Commissioners could pay off the interest on the loan from revenue. They already had sufficient funds invested to clear the outstanding balance on the current loan. Provost Maclean, seconding, said that the cost of the whole scheme would be less than half that of a Dreadnought and surely the island could claim that they had given something much more valuable than that to the Empire.

Lord Leverhulme had modestly disclaimed any influence with Government departments but painted a glowing word picture of how the earning capacity of the Lewis crofters would be increased almost fourfold by the schemes and this in its turn would contribute to the national economy in the form of increased buying power and taxes. At the end of the meeting it was agreed to ask the Lewis District Committee to associate themselves with the Commission in order to further the whole scheme of Harbour Extension and Railways in Lewis. In these last few weeks of 1918 there seems to have been a sense of optimism in the air – the war was over, the men were coming back, and there was a new, wealthy landlord, willing to pour his money into the development of the island. Peace was here and prosperity was round the corner.

Iolaire

On New Year's morning 1919 the people of Lewis and Harris woke to a hurt and grief beyond anything they had experienced before. The naval yacht *Iolaire*, bringing over 300 service men home after four years of war, ran aground and sank at the entrance to Stornoway Harbour. Through the heroism and strength of one young man, John Macleod from Ness, about 70 men were saved. Over 200 died on their home shore within sight of the familiar lights of Stornoway. There could hardly have been a crueller time or day or place to lose their returning sons and the scale of the tragedy left its mark on island life for generations.

Almost a century later it is hard to understand why not one reference to the *Iolaire* can be found in the Stornoway Pier and Harbour Minutes or the Secretary's Letter Book for January 1919. The disaster was widely reported in the national press and expressions of sympathy and donations poured in to the Town Clerk's office. Queen Alexandra, the Queen Mother, telegraphed her condolences. Provost Maclean sent the Queen Mother's message to every minister in the island so that it could be read in the churches and he was present on 10th January at the Joint Meeting of the Harbour Commission and the Lewis District Committee when, before the business was taken, the members recorded their deep sympathy with Bailie Peter Macleod and his wife on the death of their teenage son after a few days illness. Yet neither the Provost nor the Chairman of the meeting, Charles Orrock, is recorded as having mentioned the *Iolaire* disaster at this, the Commissioners first meeting of 1919. Perhaps the most likely explanation is that the Commissioners were still bound by the Admiralty restrictions on reporting of naval movements and the *Iolaire* was on naval duties. The press reported that Angus Macleod, Collector of Dues, agreed to be Secretary of the Iolaire Disaster Fund, although this is not recorded in harbour archives either. By the end of the year it totalled over £29,000. ❑

The Bigger Picture

The main subject which dominates the letters between Lord Leverhulme and the Clerk to the Commissioners in this post World War I period is, of course, the extension of pier accommodation decided upon at the beginning of 1919. There seems to have been a slight edginess in the relationship between his Lordship and Archibald Munro which comes out in frosty little comments in their correspondence.

One of the first hints of that is in a letter from the Clerk to Alexander Ross, their Edinburgh lawyer, when describing the complicated situation regarding the ownership of the Patent Slip.

'I may be wrong' he said 'but I cannot see that Provost Maclean has any valid title which the Commissioners can take over.' In another letter at the beginning of the year, he warns Alex Ross in a postscript that 'with reference to the claim that Provost Maclean has passed on to the Commissioners the rights which he acquired, I may mention that this was done verbally at a meeting of the Commission'.

In February the Clerk was telling Lord Leverhulme that 'if convenient to your Lordship I shall now be pleased to get back the plans of the proposed Harbour Extension so that steps may be taken to frame the Provisional Order. . .' He went on to say that Alexander Ross had requested terms and conditions of the sale of the Slip to Provost Maclean, but he had not been able to supply these because he had no documents.

The Leverhulme letters in the Harbour file have a slightly steely quality to them and the sharp business mind behind the genial, generous, public pronouncements can be detected in his Lordship's response.

> 'What I am concerned about is that Mr Newhall (*sic*) should be served with notice at the earliest possible moment that the Pier and Harbour Commission intend to purchase the slipway. . . I would not, however, consider that at this stage it would be advisable to inform Mr Newhall that a less portion of land might be suitable. This, of course, would come later, but if Mr Newhall retains the bulk of the land, then the payment to Mr Newhall ought to be a fraction only of the amount for which he was willing to sell the whole to Provost Maclean.'

Space for MacBrayne's

In September 1919 David Simpson of MacBrayne's wrote directly to Lord Leverhulme, complaining that his company had been pressing for adequate office accommodation for some months without success. He had heard that their latest application had been turned down, and he begged to submit that there was office space on the pier being held up by persons who did not require it.

'I had your Mr Clarke in here and pointed out to him how six Clerks, male and female, were huddled together at one small desk where there is only room for two. I find it increasingly difficult to carry out the work in these conditions and I know that were your Lordship to see them, you would not tolerate it for an hour. The health of the Clerks is suffering through this congestion and to make matters worse there are no sanitary arrangements of any kind.
I trust your Lordship can find time to give the matter consideration.'

His Lordship replied in his usual polite manner that the Commission were unable to disturb their other tenants but that it was always within the power of the tenants 'to arrange amongst themselves in some way for each others mutual convenience'.

At the Board meeting to which Mr Simpson referred, Lord Leverhulme had asked what claim MacBrayne's had on the Commission if all their other accommodation was let. A chorus of members told him 'None' and Provost Maclean pointed out that in Glasgow Messrs MacBrayne's offices were not all on the quayside and if his clerks were suffering in any way Mr Simpson should look for additional accommodation 'up town'. 'I quite agree,' his Lordship had said. He managed to weave the Provost's idea neatly into his reply to Mr Simpson, advising him the best plan would be to use the means he would have to adopt in other ports, 'Glasgow, Liverpool or elsewhere'. ❑

Archibald Munro sent on a sketch plan of the land the Commissioners thought was required at the Slip and Leverhulme acknowledged this in a note on 8th February from the Lever Brothers offices at 11 Haymarket, London, just before he left to keep an appointment with the Secretary for Scotland, Robert Munro. He left with that gentleman a copy of the draft letter which he intended to submit to the Treasury and three days later on 11th February, Mr Munro, having studied it, gave his advice. He wrote to say that it might be advisable to go into greater detail on the improvements, why they were required, how much they were likely to cost and how the repayments would be funded.

Lord Leverhulme replied on the 12th that he would have such a submission prepared, as well as a plan of the intended works. He also confessed that he had already sent one letter to the Treasury without waiting for the Secretary's advice and he had another prepared to send that day. He enclosed copies of both letters. In the first, dated 10th February, he had said that the people of Lewis were. . .

> '. . . not unmindful that free grants have been made by the Treasury for the Harbour of Wick of a quarter of a million, and free grants of lesser amounts to Peterhead, Fraserburgh and other Scottish ports. No free grant has at any time been made for the town and harbour of Stornoway. It may not be out of place to point out that Lewis has provided voluntarily a greater proportion of men to join His Majesty's sea and land forces for the war, in proportion to the population, than any part of the Empire or the United Kingdom.'

In the second, warned by Robert Munro's cautionary words, he acknowledged that he should have been more specific and sounded a quieter note. His determination and genuine wish for the good of the people still comes over strongly in the last couple of sentences.

> 'If you have any doubt about the repayment of the instalments, I should be very pleased to add my personal security to that of the Pier and Harbour Commission and the dues received by the Harbour Commission to repay the loan. I am doing this quite voluntarily as a help to the Island to secure the completion of this improvement which is essential to the well-being of the Island and without which the present poverty of the Island would continue and in the near future become accentuated.'

The same day his Lordship wrote to Archibald Munro, putting him in the picture and requesting him to have the required statement on the financial position drawn up. He stressed the fact that he, himself, would add his guarantee to the repayments. He suggested that a meeting be called and that Mr Orrock be shown the correspondence beforehand so that the matter could be fully considered. The Clerk replied that he had arranged a special meeting for 19th February and in the meantime would prepare the statement requested, but warned that it would be very difficult for him to predict future revenue which was dependent on the success of the fishing. That season's fishing had been particularly bad and if an improvement did not show that week, the figures were going to be much less than the previous year.

Meanwhile Mr Wall down in Liverpool was burning the midnight oil, preparing two plans to go with the application. He produced one which he felt was the best idea, being a solid quay wall on three sides of the basin; then another along the lines suggested by the Government engineer with an inlet to stop the waves surging during storm conditions under the floor of the wharf. Sending these down to his employer in London he referred to his Lordship's memoranda of the 10th and 12th and said that since he had only returned from London on Friday morning (this was Monday) he hoped he had not been too long in submitting these papers.

Mr Wall might have been used to his employer's breakneck speed of decision and action but the Clerk to the Stornoway Pier and Harbour Commission was not and had obviously no intention of becoming so. At their special meeting the Commissioners recorded their deep gratitude to Lord Leverhulme for his generous guarantee and set the Clerk to preparing the information required for the application to the Treasury.

Archibald Munro set out the circumstances in a letter to Alexander Ross the following day:

> 'The Harbour Commission are preparing a large scheme of Harbour Extension, the estimated cost will probably amount to £250,000, and the Chairman (Lord Leverhulme) is very insistent that this scheme, which is his own, should be proceeded with at once. . . today I have receipt of the plans of the work from him with instructions to make an application for the loan and to send it on to the Secretary for Scotland. I am doubtful of the procedure he recommends as I think that the sanction of the Board of Trade is necessary to the application for the loan and the approval of the plans. I shall be obliged if you will wire me as fully as possible, what are the preliminary steps that should be taken.'

Archibald Munro then tried to steer his Chairman into the safe waters of procedure and form. He wrote to Lord Leverhulme in London acknowledging receipt of the plans, which he said he would forward to the Scottish Secretary.

On 7th March Lord Leverhulme was back in Stornoway with an acknowledgment from the Treasury of his earlier brief application. The Lords Commissioners of His Majesty's Treasury acknowledged the application for an advance under Part 1 of the Development and Road Improvement Act 1909, and had caused it to be forwarded to the Scotch Fishery Board for transmission to the Development Commissioners in accordance with Section 4 (1) of the Act.

He chaired a special meeting of the Commission that day at which the draft of a fuller application was approved. This was sent on by the Clerk and nothing more was heard on the subject until the middle of May when Treasury Secretary's office informed them that Harbour Improvements would 'fall to be dealt with by the proposed Ministry of Ways and Communications.' Their Lordships, therefore, had deferred consideration of the scheme until after the said Ministry had been established. 'I much regret to note the decision,' Leverhulme wrote when he heard. 'It would seem to me that the reason is quite insufficient.'

After this check in his operational plans, Lord Leverhulme suddenly turned his attention to the administration of the Commission, writing to Archibald Munro on 30th May 1919 from his country residence, The Manor, at Thornton Hough, in Cheshire, to tender his resignation as Chairman of the Commission, and that of Charles Orrock as Vice-Chairman.

> 'My reason for taking this step is that I cannot agree at all with the basis on which I am appointed Chairman or Mr Orrock appointed Vice-Chairman, merely from the fact that I am the proprietor and Mr Orrock is the Chamberlain, respectively. The full efficiency of the Board requires that the Board should elect their own Chairman and Vice-Chairman from the members of the Board and I am not willing to occupy the position as Chairman merely because I happen to be the proprietor of the Island, nor is Mr Orrock agreeable to continue to occupy his position as Vice-Chairman merely because he is the Chamberlain. . . I have the greatest desire for the success and prosperity of Stornoway and it seems to me that one of the essential steps is that the Stornoway Pier and Harbour Commission should be put on a sound democratic footing, and have the power given to it of electing its own Chairman and Vice-Chairman.'

If the Clerk was surprised by this revolutionary suggestion it does not show in his reply:

> '. . . I quite see the force of the reasons assigned by your Lordship for your action, but my impression is that until there is an alteration made on the constitution of the Harbour Trust your Lordship cannot by resignation denude yourself of the Chairmanship. I would suggest that you consult your Edinburgh Agents on this point.'

As requested he put Lord Leverhulme's letter before the

Commissioners and they instructed him to tell his Lordship that while they fully appreciated his democratic reasons, they sincerely hoped he would reconsider.

Lord Leverhulme was by this time back at Lews Castle and he wrote to Archibald Munro at his home in 20 Bayhead Street asking for a meeting to be called on the following Tuesday morning 'and you could lay the matter before me, as I am convinced that the course I am taking is in the best interests of all concerned.' When they met the Commissioners heard an apology from ex-Provost Anderson explaining that owing to a severe accident which had befallen him, he could not be present but he hoped his Lordship would continue the Chairmanship. The Clerk then read the resignation letter again, before going on to read the relevant clauses in the Stornoway Harbour Order 1865 in which it was declared 'that the Superior or his Chamberlain of the time should be permanent members of the Commission, the one in the absence of the other, but the one only in the absence of the other shall be when present, Chairman.'

His Lordship then gave a statement of his reasons for resigning. He began with the worthy thought that he knew of no other Public Board which did not have the right to elect its own Chairman, who should have the full confidence of its members. He could not continue to act as Chairman unless he had the full confidence of all the Commissioners and unless he was kept informed of any important business which was to be considered by the Board whether he was in residence at the Castle or not. The Chairman as a rule was communicated with in his absence and he had not found that he had been so communicated with, for example in the recent appointment of Captain Macdonald to the important post of Harbour Engineer. (Rae Macdonald, taking over from his late father, Alexander)

Lord Leverhulme was either unclear as to the status of his Chamberlain in his absence, or else he was using the situation to make a point. The *Stornoway Gazette* quotes the Clerk as stating that he and his predecessor had looked on the Chamberlain as Chairman of the Commission in the absence of the Superior and that what was done with the Chamberlain in the Chair was done as in the presence of the proprietor. Kenneth Mackenzie explained that the meeting had been called for at the request of four members of the Commission addressed to the Chairman. He assured Lord Leverhulme that the whole thing was a misunderstanding.

The Commissioners apologised to his Lordship and assured him that Clerk had been following the usual procedure and that the omission to inform him had nothing to do with a lack of confidence in him. Each Commissioner present expressed their entire and full confidence in him as Chairman of the Commission and they unanimously repeated their request to reconsider. Naturally, he did and withdrew his resignation, which action was met with enthusiastic applause.

The Minute of the meeting ends with a sentence directed at Archibald Munro:

> 'It was an instruction to the Clerk to supply his Lordship, if outwith the island, with timeous notice of any business of any importance to be dealt with at Commission meetings. . .'

Which was Munro-speak for Lord Leverhulme's expressed wish to be kept in close touch with the business of the Board so that he could communicate his own views, even by telegram. His Lordship was back in the driving seat.

The correspondence between the two men becomes very brief for some time after this.

Lord Leverhulme schemes were not receiving the whole-hearted welcome he had expected. While he had been preparing his application for funds in March, the farms at Gress and Coll had been raided by men from the Back district, who, tired of waiting and talking, had marked out crofts for themselves which they intended to settle. This had been followed in May by the raiding of Galson Farm. His Lordship was perplexed by his inability to bring the raiders round to his plans, which, as it seemed to him, would give them such a better way of life. In the week when he had received the Commissioners' assurances of confidence, he had been in conflict with the raiders again and in no mood to compromise. *The Scotsman* pointed out that Lord Leverhulme 'lacks the full confidence of the people of the island at the present moment' and that the crofters had only a hazy notion of what he really wanted to do.

There was unease in other sections of the community as well as the crofting population. In a Town Council meeting in March, Angus Macleod, the Collector of Dues, who was also a Councillor, moved that they ask the Pier and Harbour Commission to defer applying for a Provisional Order until the Council and the community had had an opportunity of discussing the proposal. He said that although they had three representatives on the Commission, the Town Council had only heard of the proposal from the local press. Provost Maclean said he saw no seconder and suggested they drop the matter but this roused Councillor George Stewart to second the motion in order to keep the discussion alive.

Angus Macleod's views seemed to have some support in the town, judging from the reaction at a Ratepayers meeting in April called to discuss the town water supply. A suggestion from John Macritchie Morrison that Lord Leverhulme's engineer should be asked for his advice was met by booing and cries of 'No, no!' and 'Too much Leverhulme!'

Bailie Roderick Smith, one of the Council representatives on the Harbour Commission, offered to stand down when the matter of the annual representation came up again on account, he said, of a general feeling expressed that he and colleagues had not done their duty on the Commission. Peter Macleod assured him that there was no reason to stand down and that he was in good company when criticised. Bailie Smith said he was referring to a general feeling which prevailed in the community, he did not object to being criticised.

Over the first twelve months of his proprietorship, Lord Leverhulme had bought the Stornoway Gasworks, bought the British Oil and Guano Company, bought over the controlling interest in Kenneth Mackenzie's carding mill, bought over the Newcastle company which produced Skipper sardines; bought over one hundred fish shops 'in London and the provinces', and bought North Harris from Sir Samuel Scott. He was spending thousands of pounds in wage bills every week on the upgrading of some of the roads in the island, and on the new road from Bayhead to Arnish. There was talk of a fleet of 10 or 12 oil-fuelled trawlers fishing out of Stornoway to the waters of Iceland and the Faroe Islands. Every school in the island had received a copy of his map showing the port at the centre of huge red circles at 50 mile intervals up to a radius of 750 miles. He intended to organise the Lewis fishings, he told a *Scotsman* correspondent that June.

> 'My proposals are to link the whole thing up. The men will catch the fish, the girls will prepare them for table, my ships will carry them to the markets, and my shops will sell them. I am not opposed to crofting. . . I don't want to take them away from the land if they are unwilling to come. It would be like tearing them up at the roots.'

Throughout most of 1919 he flitted backwards and forwards between Lewis and his mainland homes, pushing on his schemes. In March he submitted to the Commissioners the plan of a new road from the foot of Kenneth Street across the Bayhead River to Arnish to which they offered no objection. In July he chaired a Harbour Commission meeting at which it was decided to make representations to Robert Munro, the Secretary for Scotland, on the subject of the loan. The Clerk was instructed to draw up the letter and submit it to his Lordship for revisal. Archibald Munro's draft is still in the harbour file, with its red pencil revisions as it came back from Lews Castle on 27th July.

On the 29th his Lordship wrote from Hampstead about lighting the Harbour with electricity. On the 4th August Archibald Munro was sending on a copy of the old Provisional Orders to Port Sunlight, and a reply dated 7th August from Hampstead states that his Lordship is leaving for Stornoway the following day. His social diary can be traced through press reports – Lord Leverhulme lecturing at the UF English Church on the subject of 'Sympathy'; a dinner for Lord Leverhulme in the Imperial Hotel hosted by a number of prominent local businessmen; Lord Leverhulme addressing the Philosophical Institution in Edinburgh; a reception for Lord Leverhulme at the Lewis and Harris Gathering in Glasgow.

At the end of September the Commissioners agreed to send a deputation to see the Secretary for Scotland and Alexander Ross was asked to draw up a Memorandum for submission at this interview. Before this was completed, however, the Scottish Secretary himself contacted the Clerk to say that the Fishery Board Engineer had estimated the probable cost of the harbour works at £1 million. Archibald Munro immediately wired Port Sunlight with this bombshell and Lord Leverhulme replied that he should write to the Scottish Office and ask for full particulars of how the figures were arrived at.

> 'I am of the opinion that a solution of the difficulty may be made by constructing the work in sections We are bound to have a reduction in cost of such works each year as conditions of labour return nearer normal. At the same time, it would be rather disastrous to hang up the whole scheme indefinitely. My solution would be to do as little as possible at present, but to do something, so that we could get a start on fishing developments at Stornoway.'

The Secretary for Scotland informed the Commissioners that he felt a deputation would serve no useful purpose at the moment. Before leaving on a trip overseas, however, Lord Leverhulme chaired the October meeting of the Commission in which he complained that unless something was done for Stornoway, it was going to go back rather than progress. He was going to develop the Harris port of Obbe (present-day Leverburgh) at his own expense. He would not wait for the Scottish Office who would take ten years to get round to anything. Because there was a Harbour Commission in Stornoway, he was not able to do in the port what he could in Obbe. He had also requested that a financial statement be drawn up and pointed to the fact that they were not collecting enough revenue to meet expenses. Councillor Angus Mackenzie seconded his Lordship's motion that all the rates in the Schedule be increased to their limit under the existing Provisional Order and the Commissioners unanimously agreed.

The division in the community over the Leverhulme plans was an added sadness in 1919, a year which had seen such turmoil and heartbreak. On 4th August the town had joined in the Peace Celebrations, a day of thanksgiving and laughter and music which began with a religious service in the Square, continued with marches, a fancy dress parade and a concert, and ended with a display of fireworks from No 2 Pier and Goat Island, organised by John Macritchie Morrison. Flags had fluttered from every house and between the Castle turrets, and in the evening both the castle and the houses all along the seafront had been illuminated.

In contrast, the very first Armistice Remembrance Day was observed on 11th November, when at precisely 11 am the Red Ensign was run up the Harbour flagpole and most of the steam drifters signalled the hour to the town with two short blasts on their sirens. Everything immediately fell silent, the unloading of cargoes stopped, the bustle of the harbour stilled, and the only sound was the restlessness of a horse startled by the sirens. The *Highland News* reported a little scene in one of the shops where a few women were shopping:

> 'The merchant said 'It's now eleven o'clock. We will just wait for two minutes by the King's wish.' Nothing further was said but ere the two minutes elapsed the little group of customers were all in tears.'

No Progress

The large developments on which so many hopes were pinned began to look less likely. Lord Leverhulme chaired a meeting of the Commission on 9th February1920 and outlined what he intended to do next. The gist of his speech was that a) because of the raiding of one his farms in the rural district of Lewis, he would concentrate his work on Stornoway and Harris; b) in view of the Government's refusal to help the harbour development scheme and in order to get on with the necessary development, he intended to carry out, at his own cost, the very minimum of work required on this scheme and had had Mr Wall prepare a plan for the Commissioners' approval; c) the total revenue of the harbour would be applied first to the existing liabilities and working expenses and then any balance remaining he would expect the Commissioners to make over to him towards refunding the cost; and d) he would not ask for any security on the money on the assumption that the future prosperity of the port would be enable them to pay it off at a later date. He was insistent that, although he would carry out the minimum works required, the Commissioners should press the Government to do something for Stornoway commensurate with what they had done for other Scottish ports. If the Government did come in on the scheme, he said, he expected to be treated *pari passu* with them in the repayment of any capital.

The plan he proposed, which received the Commissioners' full approval, was to connect Goat Island to Newton by means of a breastwork and to erect a jetty on the island. The Commissioners also resolved to ask for powers in their proposed Provisional Order to allow Lord Leverhulme or his successors to be paid out of surplus revenue, and unanimously accepted his offer. Archibald Munro sent a copy of the Resolution to Lews Castle but his Lordship wrote back immediately to correct what he thought might be a wrong assumption on the Commissioners' part.

> 'There is one point which has been omitted and to which I attach the greatest importance, which is that the Goat Island should remain the property of the Estate. It is essential that I should provide buildings for the fishing industry proposed and some of these buildings would require to be erected on Goat Island but if the Island were feued by the Harbour Commission or at any later date an attempt were made to cause it to be feued, this then would defeat the object I have in view, namely to obtain facilities . . . at the earliest possible moment rather than with two or more years' delay. . .'

He added that of course the Harbour Commission would have the right to levy dues at this proposed jetty and he wished no exceptional treatment on that subject, but he did want security of ownership on Goat Island as a very essential condition of his financing the construction works. Hume McGregor, his Lordship's lawyers, also weighed in, writing to Archibald Munro that they had drawn a new version of the Resolution which they felt met the case, 'as the wording of the Resolution you had framed looked somewhat as if Lord Leverhulme was making a present of the Pier and Breastwork to the Harbour Commissioners.'

Archibald Munro, however, had been doing his own homework. He had contacted Lord Leverhulme himself in a letter to Hampstead Heath on 14th February 1920:

> 'Might I venture in my private capacity to ask your Lordship if the Quay Wall to be built round and about Goat Island would come under the reservations which your Lordship stipulates with regard to Goat Island? I tender my apology for troubling you in the matter.'

His Lordship sent a full answer, re-iterating that of course the Quay Wall would remain the property of the Estate but that vessels landing cargo there would be required to pay the same dues as at Commission quays 'but if the wall surrounding Goat Island and which were necessary to form a harbour do not form part of the Island and part of the Estate, then there would be no access by water without the permission of the Harbour Commission and the money expended by the Estate would have ceased to be part of the Estate.' Naturally, he ended, whenever the property was paid for, it would revert to the Harbour Commission but it had to remain the private property of the Estate until fully paid for.

The Clerk replied with unusual tact:

> 'What I fear, speaking for myself, is that the Harbour Commissioners in asking for powers under the proposed Provisional Order to pay for and maintain the proposed works as shown on the plan submitted, out of surplus revenue, that such powers might not be granted without the Commissioners having a title to the subjects, that being my opinion in view of the Commissioners' experience in connection with their previous undertakings. I should like to have your Lordship's guidance and opinion on this point.
> Your Lordship's obedient Servant. . .

A fortnight later he was writing to Hume McGregor with a copy of this letter with the comment: 'I got no enlightenment from his Lordship further than referring me to your amended Resolution. Can you tell me what is the actual position of matters?'

The Minutes of the Annual Statutory Meeting record for the first time a Commissioner nominated by the Lewis and Harris Welfare and Development Company Limited, 30 Saint James Square, London, Proprietors of the Island of Lewis and Superiors of the Burgh of Stornoway. Kenneth Mackenzie represented the new ownership of the Island which had been made over to the Company by his Lordship in early December 1919. When Archibald Munro came to know of this action is not clear, since the first reference to the company in the Harbour archives comes in a letter to him from Hume McGregor dated 22nd March 1920. The lawyer announces he will be in Stornoway in a few days and hopes the Resolution on the Harbour works has now been passed. He adds that he could have a letter signed by Lord Leverhulme's Attorney or one 'signed by the Directors of LAHWAD undertaking any obligations undertaken by Lord Leverhulme', giving the impression that the Clerk is familiar with the situation.

The newspapers of the time give virtually no prominence to the establishment of this Company although the launching of Leverhulme's commercial enterprises were usually very fully reported. There was an international slump in the Spring of 1920 and Leverhulme was spending huge sums of money in an enterprise which perhaps he was beginning to realise was not as certain of success as he had believed. He had been locked in combat with the farm raiders since his return from America, sacking about 60 men from his development schemes because they came from the districts of Back, Coll and Vatisker and announcing they would be reinstated if the raiders gave up the land they had taken there. If he hoped that public opinion would cause the raiders to pull back, he was wrong.

Opinion was certainly divided – *The Scotsman* reported in March that there was a feeling of tension throughout the island and that a petition had been signed by over 7000 Lewisfolk demanding he take legal action

Hurricane

On the morning of Wednesday 16th March 1921 Lewis was hit by a storm of hurricane force, with gusts reaching over 90 miles an hour. An anenometer placed at the Battery recorded a gust of 87 miles an hour before it was blown away. In the two or three hours that it raged the storm claimed four lives in Point, when a house in Aignish was blown down and caused huge damage throughout the island. Masonry from the ruin of the Town Hall crashed down into the Chamberlain's office which had been rebuilt within the walls, Estate staff narrowly escaping by taking shelter in the passage through to Carn House next door.

At the Fish Mart five heavy brick chimney stacks fell through the roof, fortunately causing no injury. In the Inner Harbour some boats were lifted out of the water and deposited on the quay while on South Beach a boat of around 20 ft keel was thrown from No 2 Pier onto Cromwell Street. This particular vessel, according to the *Highland News* was no light craft but was used for transporting sheep on and off the Shiant Islands. Further along South Beach, towards No 1 Pier, a small boat was impaled on the railings at Tigh-na-Cloinne.

The Works Committee of the Pier Commission met the following day to review the list of damage drawn up by Captain Watt:

Fish Mart	Badly damaged
Cargo Store at back of Antrim Co's store	Considerable damage
Back Store at Messrs MacBrayne's	Damaged
16 Quay Lamps	Damaged
Wooden Hut at No 2 Wharf	Demolished
Roadway between No 2 and SE end at Slip	Considerable damage
Landing slip at No 3 Wharf	Slight damage

In view of the urgency, they remitted to the Works Convener, the Clerk and the Harbour Master to get offers from contractors on a percentage basis by 7 pm the following day. James Macrae's offer at 10% on journeyman's wages and any new material at invoice prices plus 10% was accepted. The Commissioners also agreed to lower the chimneys on the off and left side of the Fish Mart. ❑

against the raiders – but it did not prevail. Most of the raiders remained on the land and faced the consequences. An injunction was taken out against them but his Lordship did not seek to have it served. Instead he ordered the suspension of all his works in Lewis in the hope that this would have more effect.

He was not in Stornoway on the 7th May when the decision was taken to stop work on all his schemes, an action which left hundreds of men out of work at a stroke. The new roads were abandoned and the Commissioners' new harbour scheme was shelved, a note from Leverhulme to the Clerk saying that he was afraid the present position in the Island of Lewis would prevent him from proceeding further in the matter.

The public bodies were thrown into consternation and appeals were made to the Secretary for Scotland to formulate some policy on land settlement that would put an end to this impasse. A huge and fiery public meeting in the Drill Hall condemned the raids and appealed to Lord Leverhulme to reconsider.

When Lord Leverhulme did return to the island on 5th June he received a magnificent welcome, with thousands of people turning out to meet him and cheering him enthusiastically. In the Harbour Commission meeting, a few days later, he made a statement on his decision, and the Commissioners 'expressed their deep regret that the works had been suspended on account of the lawlessness of a few persons and the action of the legal authorities in connection therewith'. His Lordship thanked the Meeting for their expression of confidence and there being no other business the meeting was closed.

No Settlements

Throughout 1921 the Commission's deliberations and decisions were all made against the background of Lord Leverhulme's threats not to go on with his schemes. In the early part of the year it had seemed as if things were back on track, with assurances from the Secretary for Scotland that land settlement would only be on the land which Leverhulme had made available and no compulsory purchase of any land would take place while his schemes went on. This elicited the response from his Lordship that he could now restart work on some of the works, among which was the extension of Stornoway Harbour.

'When this harbour is complete, Stornoway will possess the finest fishing harbour on the West Coast and Western Highlands of Scotland,' he predicted in January. However in early May, some ploughed land at Coll Farm was sown by men from the district and consequently Colonel Lindsay carried out Leverhulme's instructions and paid off the men who had only shortly before been taken back on to the harbour works and the building of the new Tolsta Road. According to *The Scotsman* the island had never before been 'in such sore straits. The fishing industry is practically at a standstill and the tweed industry stagnant.'

The schemes had remained suspended throughout the year and Lord Leverhulme, distributing the prizes at the Stornoway Highland Games at the end of August, used the occasion to underline that they would remain so for the foreseeable future. This decision was taken on mainly economic grounds, shaped by the prevailing worldwide depression. His Lordship reminded his listeners that from the very beginning he and the people had been agreed 'that our relations would be on a strictly business basis, that there be no odious taint of philanthropy to lower yourselves in each other's esteem.'

He blamed the slump in world markets for the fact that the canning and ice factories could not be opened and advocated patience until more favourable conditions returned. By November 1921 the Government, faced with the continued reports of poverty and suffering coming from the islands, felt that they could not keep their previous promise of giving him ten years to prove his schemes. Lord Leverhulme had not restarted the works as he had promised and, in the circumstances, the Scottish Secretary did not feel the Government could be held to their undertaking. The farms of Coll and Gress were therefore broken up into small holdings for the people of the district and although Leverhulme did not challenge the decision, he voiced his disapproval in the national press and asked if he intended to abandon his projects, replied that they were merely postponed. Asked to put a date on that, he answered, 'If you can tell me when things will get normal again, I will.'

Things were not going to be normal for a considerable time to come. The Commissioners who lobbied the Secretary for Scotland for grant money in early December 1921 were Provost Roderick Smith, Councillor John Macritchie Morrison and Bailie Kenneth Mackenzie. With their Town Council and Lewis District Committee hats on, they were part of a deputation to the Government on the woeful mail service given to the islands which they contended had been set back fifty years. As Roderick Smith pointed out to the Secretary for Scotland a letter from London could be delivered as quickly in New York as it could be delivered in the rural parts of Lewis under the existing conditions. The trio were accompanied by ex-Provost John Mackenzie, W. C Mackenzie, the historian and Roderick Macleod, another prominent London Lewisman, as well as Sir Herbert Morgan and Sir Edgar Sanders, representing Lord Leverhulme. They outlined the inadequacy of the service, the ruinous effect it had on trade and the appalling conditions in which the fishergirls had been travelling over the winter. The

Chancellor was courteous and sympathetic but warned that the cost of the war had placed heavy burdens on the Exchequer and he could make no promises.

The winter of 1921-1922 brought no improvement to the transport or communication situations. During the usual January gales the island's telegraph cable was again cut off and the GPO set up a temporary wireless connection until the repairs could be effected. The winter fishing trade was badly hit by the lack of communication. In April nine prominent citizens wrote to *The Scotsman* detailing a hair-raising journey they had just had on the steamer *Gael* which had broken down for about two hours while making the crossing. The previous week the *Sheila's* pumps had refused to work and that vessel had to return 'partially waterlogged' to Stornoway. Although both vessels were now equipped with wireless, the station on the island was no longer worked and the letter asked what would happen on a stormy night in the event of similar breakdowns at sea.

The Commissioners began 1922 with the intention of getting the improvements to the Inner Harbour underway as soon as possible. To this end they engaged the services of Archibald Henderson, an Aberdeen Civil Engineer recommended by Gordon Nicoll, the Fishery Board Engineer. This was the first contact with a firm which would carry out some of the Commission's major works down through the years, right up to the time of writing. In January 1922, however, Mr Henderson made his first visit to the Inner Harbour accompanied by the new Chamberlain Captain Fletcher, Kenneth Mackenzie, the Harbourmaster and the Clerk, then returned to Aberdeen to draw up a report, which he submitted two months later. The Commissioners decided to apply to the Scottish Fishery Board for grant and loan in order to carry out the first instalment of his recommendations. Kenneth Mackenzie had a personal interview with the Chairman of the Fishery Board who advised them to make the application to the Development Commissioners. Mr Jones also advised that the Annual Accounts should be sent to the Treasury and that Hume MacGregor should be appointed to carry out the Provisional Order which would authorise the improvements In May 1922 the Commissioners asked Henderson to add an appendix to his report showing an extension to Esplanade Quay and then the matter stalled.

'What is their standing?'

Part of the reason for the delay was financial. That month the Commissioners received a letter from the Head Office of the National Bank of Scotland warning them that before their 'indebtedness to the Bank' was increased and after the War Loan of £3,500 was applied, they would have to put a fresh request before the Bank along with last balance sheet. On no account, said the Bank, should the money received from the sale of the War Bond be put to other expenses. The Finance Committee recommended that the Bank should be asked to accept a Minute for £8000 and the balance sheet should be accompanied by a full explanation of the recent expenditure on repairs. Archibald Munro did his best, pleading that the expenditure of £13,000 over the last two years had considerably enhanced the value of the property. As the matter was urgent he asked that the Bank's reply be telegraphed to him. Angus Cameron, the local agent of the National Bank, sent them a copy of a letter to himself signed by A N Macleod for the Secretary of the Bank in Edinburgh.

The letter said that the Bank had given weight to Angus Cameron's recommendation that they should sanction the £4,270 due by the Harbour Commissioners for Local Rates but they were unable to advance any further money without security. They had considered the application sympathetically 'realising the importance of the matter and its bearing on the relationship between the Branch and perhaps important interests in the town' but they could not overlook 'in our responsibility as lenders the present position in which the Commissioners find themselves'.

The Revenue was not sufficient to meet the ordinary expenditure and unless there was an improvement in the trade of the port, the deficit would increase. The Bank was therefore obliged to look for security from the value of the assets but they required to know what these were and their value. 'We should also wish to see a copy of the Constitution of the Commission for our information. Who are the Commissioners and what is their standing locally?' They added that they might also need personal security from the members.

The Commissioners replied that they could not entertain the giving of any personal security since they considered the £80,000 value of the Harbour property to be ample security and they would like to know what security was required. Archibald Munro wrote accordingly, listing the ten Commissioners, starting with the Chamberlain of the Lews. He also, as instructed, checked with Hume MacGregor whether it would be in order to give the Bank part or whole of the Harbour Property as security and whether that would prejudice their application for grant and loan from the Development Commissioners. The Bank replied that they had reconsidered and were willing to authorise a working credit for the £8000 with a security on their property as well as an Assignation to the Commissioners' powers to levy dues. This was to be kept to the strict limit and for a year only.

'I may mention,' Angus Cameron concluded, 'that the Directors did not view the proposition on its actual merits, with any great favour, but they desire as far as possible to extend consideration to customers whose connection with the Branch is held in high esteem.' After discussion the Commissioners decided to offer the Bank as security the dues and rates earned over what was required for ordinary expenditure. This offer was made on the advice of Hume McGregor who had also recommended that Counsel's Opinion should be got before they did anything.

The Bank refused to move on their demands for security on property and the Commissioners delayed a decision for the moment. The Clerk was instructed to ask the Public Works Loan Board what documents were given to the Board when the loan of £30,000 was granted in 1892. He had already confessed to Hume McGregor that he could find no such documents among the papers or in the Minute Books. When the answer came through that the Loan Board's only document was a mortgage securing the loan, and the Commissioners found themselves again liable to meet the half year's instalment due on it, they took the decision to reduce the overdraft amount requested to £5,000.

Archibald Munro, reporting this to Angus Cameron on 8th August, was able to note that 'the circumstances of the Harbour have considerably improved for the last few weeks owing to the good fishing which has set in and in view of a continuance of the same the Commissioners consider that your Directors should have no hesitation in granting the advance.' It was September before the agreement was made, the Commissioners putting up the Dues and Rates of the Harbour as security, subject to the prior charge of the Public Works Loan Board.

Who are the Commissioners?

Another reason for delay in beginning the Inner Harbour work was the decision to promote a new Provisional Order and a Special Committee consisting of the whole Commission had been set up to deal with this subject. One of their first recommendations was to include powers to construct the extension to the Outer Harbour as designed by Mr Wall of Lever Brothers. Captain Fletcher was also asked to find out if Lord Leverhulme would agree that the Order could be amended in such a way that the Commissioners would be able to appoint their own Chairman.

Provost Smith raised the question of the qualifications required by the Town Council representatives, citing the difficulties experienced in getting three members from the Council ranks who satisfied the existing rules. He also wanted them to look at the qualifications needed by both the Duespayers and those they elected. Provost Smith first brought up the idea at the Committee held on the morning of 8th May 1922 but no recommendation was made and the meeting was adjourned. However later in the day the Town Council met for their regular monthly meeting and the Provost raised the matter there as well. He pointed out that at present the Council could only appoint those of their members who were householders and proprietors to the value of £10 per annum and on at least one occasion they could only find two Councillors who met the qualification and their representation was one member short for about six months. At the time it was felt that the franchise should be extended to allow Councillors with a residential qualification to be appointed to the Harbour Commission and now that a new Provisional Order was being drawn up, the Provost felt the Council should make some representations on the matter.

Councillor A J Murray made a motion to that effect and George Stewart voiced his support but surprisingly the proposal met with some opposition. Louis Bain said that the matter had been sprung on him and he had been thinking it over all day since the Harbour Commission meeting in the morning. He found he could not agree, giving as his reason that the harbour was the biggest paying concern they had in the burgh and that it was possible that people with little or no interest in the town or the harbour might become Town Councillors. It would be a pity if such people were appointed to the Harbour Commission, he said, pointing out that there were at least five or six Councillors round the table that night who held all the necessary qualifications. Provost Smith said that Councillor Bain must know of the difficulties the Council had experienced since 'he was himself the party who came to the rescue by being in the first place co-opted into the Council and then being appointed to fill the vacancy in the Harbour Commission.' Councillor Bain replied that what surprised him was that the matter had come up before the Harbour Commissioners that morning and he had not expected it to come up so soon in their deliberations on the Provisional Order. He thought they 'would have taken three or four days on that job'.

The Provost reminded the meeting that no one had disagreed with the widening of the franchise when it had come up before in the Council and that he did not believe any future Town Council would appoint anyone to the Commission who did not have an interest in the harbour. Roderick Smith was determined that the Council should take the opportunity to safeguard its representation on the Harbour Commission and he warned that the same difficulties might arise in the future if they allowed the situation to go by default. George Stewart, backing him up, said he thought it was impossible that anyone without an interest in the town or harbour would be elected to the Council. Louis Bain, however, said that 'without mentioning any names' he thought Councillor Stewart would agree there had already been Councillors who had no interest in the town and ran it into a great deal of expense. Councillor Stewart, obviously aware who he meant, protested that 'The member you refer to had a very keen interest in the affairs of the town'.

Councillor Murray put an end to this exchange by rewording his motion to say that the Council ask the Harbour Commissioners to change their new Provisional Order to accept as representatives from the Town Council any of its members without any further qualification. He was seconded by Bailie Hugh Macleod who stated that in these days when they had a broadened Franchise in every Local Authority, Public Board and in Parliament, there was no reason why they should stand upon any old order with the Harbour Commission. No doubt the people who framed the existing Harbour Order had good reasons to put in these qualifications at the time but they had gone a long way since that Order was formed and if they had a democratic electorate in one authority there was no reason why they couldn't extend it to another. In his opinion anyone who was fitted to be a Councillor was fit to sit on the Harbour Authority. With that the discussion ended and the motion was agreed to unanimously.

The adjourned meeting of the Harbour Committee was resumed a week later and the Town Council's proposal was rejected. Captain Fletcher moved that 'that part of the Order of 1865 dealing with the qualifications of Town Councillors to be members of the Harbour Commission stand at present.' Norman Maciver seconded. Roderick Smith moved an amendment to the effect that Town Councillors qualifications be the same as those required by the Duespayers' representatives. He was seconded by George Stewart. but defeated by 5 votes to 3. At the same meeting George Stewart's motion to change the Duespayers' voting entitlement was unanimously agreed, making the qualification the payment of £2 or upwards in Harbour Dues in the financial year immediately preceeding the election.

This response put the Councillors' backs up. Reporting to the Council in June 1922, the Provost said that the Harbour Commission had been created 'largely through the instrumentality of members of the then existing Town Authority. Although Sir James Matheson made it a condition that feuars only could represent the Town Council, the whole position has changed materially in the meantime through the widening of the franchise for practically all public bodies.' Mr Smith said that he felt the Harbour Commissioners had turned down the proposal in rather an off-hand fashion. The only argument offered against the idea had been that every representative on the Commission ought to have a stake in the community.

'Well,' said the Provost, 'I think we all have a stake in the community and members of the Pier and Harbour Commission are not more interested in, nor have they greater enthusiasm in regard to the town than the members of the Town Council have, however small our stake in the community.' Councillor Murray, expressing his disappointment at the attitude of the Commissioners, suggested that instead of asking that any three members be accepted, they ask that any three members being householders of a rental of £10 to £15 might be accepted. Councillor Bain said he had suggested something along those lines during the discussion in the Harbour meeting but it had been turned down. However, the Councillors went ahead with Mr Murray's suggestion, only changing it to a definite figure of £15 annual rental.

In August Archibald Munro wrote to himself as Town Clerk, stating that the Commission had turned down this suggestion and had decided by a majority to abide by their previous decision. The discussion had been fairly heated from the beginning. When Kenneth Mackenzie was just about to move the Council's proposal Councillor John Macritchie Morrison intervened to remind the Commissioners that this matter had been settled and confirmed in previous meetings and the decision had not been rescinded. Therefore he said he thought the motion was out of order.

Captain Fletcher replied that this motion was in response to a further application from the Town Council and ex-Provost Maclean said the proper procedure would be to rescind the first decision. Bailie Mackenzie moved the proposal as suggested by the Council but said that he had only decided to do so after 'the most serious thought' and with the desire to bring this matter to an amicable settlement. Councillor Louis Bain seconded him but Councillor John Macritchie Morrison, also saying he had given it serious consideration, moved the previous question. He believed that, since the harbour was the best asset of the town, the Commissioners should be men who had a stake in the community and it would be a great mistake to alter the qualification. Ex-Provost Murdo Maclean seconded, echoing Bailie Mackenzie' views, and adding that irresponsible people should not sit round that table. John Macritchie Morrison's amendment for the status quo was carried, after discussion, by four votes to three.

Provost Smith and his supporters did not give up. In October the Clerk reported that he had given all the Commissioners seven days notice of the Provost's intention to move the rescinding of all the Minutes dealing with the matter and to put forward again the Council's proposals but the Provost himself moved that the motion lie on the table until their next monthly meeting, because there was a possibility of a Town Council election. Consideration of the motion was postponed at their December meeting and it was January 1923 before it was finally put to the vote, by which time Kenneth Mackenzie had been elected as Provost of the Burgh in place of Roderick Smith. The ex-Provost made his motion and was seconded by George Stewart, producing an extract from the Town Council Minutes authorising their representatives to continue acting on the lines already agreed in Council. John Macritchie

Morrison and Murdo Maclean again moved and seconded the previous question. They claimed that the matter had been discussed and decided more than once and that if decisions were to be questioned and overturned every time a member disagreed, then business would become untenable. John Macritchie Morrison said that if the Town Council had a mandate from the ratepayers then he would not oppose it but they did not. Ex-Provost Smith pointed out that, if there had been a contest at the recent Council election, then the matter would have come to a ratepayers' meeting. 'We are here only as mouthpieces of the Council,' he said.

However, John Macritchie Morrison's amendment again won the day, after discussion, by one vote. Bailie George Stewart moved that when the Draft of the Provisional Order was sent to Hume McGregor in Edinburgh, the lawyers be informed of the difference of opinion of the Commissioners on this issue. He was seconded by ex-Provost Smith and again opposed by John M. Morrison and Murdo Maclean, both of them intervening before Captain Fletcher could say anything.

'If the Chairman says I am out of orderI will submit to the Chairman's ruling, but there seems to be two or three Chairmen here,' protested Bailie Stewart. After another few minutes of confusion, Captain Fletcher admitted that perhaps he had been wrong to allow Bailie Stewart's motion and said that he considered they had been wasting two much valuable time and retarding the most important work of the Commission on a very small matter. 'Principle has never been a small matter,' said Ex-Provost Smith, 'and principle is at stake here.'

The vote was an even four to four, Norman Maciver abstaining and openly admitting he was sitting on the fence, and Captain Fletcher gave his casting vote in favour of the status quo. For the moment the Councillor's qualifications remained as set by the Order of 1865 and Archibald Munro informed himself of the decision in his capacity as Town Clerk.

A Fair Record

This division between the members was not confined to the question of the Provisional Order. It had also occurred, if not actually begun, in an argument at the beginning of the year about that hardy perennial subject, the Harbour Collector's salary. The unfortunate thing was that Angus Macleod had been ill since the middle of September 1921 and had been on full salary since then. Archibald Munro wrote to him on 7th March 1922 to say that 'the Commissioners all expressed their deep sympathy with you in your protracted illness' but that on the advice of the Finance Committee they had decided to pay him his full salary for March and then 50% of it for the next two months and if he was still not well enough to return to work they would have to consider filling up the post. 'I may say,' the Clerk added, 'that all the Commissioners would be glad to see you restored to such a condition of health that you may be able to resume your duties at an early period.'

The Clerk's kind words were countered by the full report of the meeting appearing in the *Highland News* the following week, in detail that would not be permitted today. It showed that Councillor George Stewart had dissented from the recommendation of the Finance Committee and reported his explanation that he had done so on the grounds that it was too drastic a cut and 'a very harsh method of dealing with a servant who has served us efficiently and faithfully for so many years.'

He had suggested the Collector receive full pay for six months and then 75% for three. His closing comment was that too much time was spent haggling over little matters when thousands of pounds expenditure got 'scant attention'. John Macritchie Morrison had taken exception to that remark and Ex-Provost Murdo Maclean, stating that there was no point in having sub-committees if all their recommendations were to be thrown out, had moved the Finance Committee's report and recommendations be approved.

Provost Smith had then joined the debate. He had not been present at the Committee meeting which had made the recommendation but he thought he saw a way out of the problem without the Commission dividing on it. It was a rather unfortunate case and they knew the conditions that had previously held. 'Possibly Mr Macleod did not always see eye to eye with his employers and at times there were difficulties with the management of the Collector's department. But circumstances are now entirely changed. I would suggest that he be paid his full salary for six months.'

Kenneth Mackenzie had replied that this was exactly what they were doing. The Collector had had his salary for five months already and the recommendation was to give him another full month and then half for two months more. Provost Smith had asked if the matter would come up for consideration at the end of these two months but Captain Fletcher, as Chairman, stepped in to ask if the report and recommendations were accepted. John M Morrison seconded Murdo Maclean's motion and Councillor Alexander Maclennan said he was prepared to support it providing the matter came back for consideration after two months.

George Stewart moved an amendment that the recommendation be adopted with the exception that instead of 50 per cent 75 per cent be substituted and Roderick Smith seconded. John M. Morrison contended that the recommendation should either be approved or returned to the Finance Committee for further consideration. 'Or modified,' observed Councillor Maclennan.

'I don't think you can modify the terms of the report,' persisted Councillor Morrison.

'That's for the Chairman to decide,' replied Councillor Stewart. The Chairman agreeing with John M Morrison's view, Councillor Stewart stated that he would therefore move that the matter go back to the Finance Committee. 'At the last meeting there were only three present – the Chairman, Bailie Mackenzie and myself, and of the three one was a dissenting party.'

Bailie Mackenzie bristled that he declined to be a member of a Finance Committee to discuss this matter again but Provost Smith seconded the amendment. The matter went to the vote and the *Highland News* reported that the amendment received three votes, Councillor Maclennan siding with the mover and seconder, while the other six Commissioners present voted to approve the recommendations. The report of the proceedings was printed on the page given over to Stornoway news and appeared opposite an obituary for one of Angus Macleod's sisters.

Three weeks later on 22nd March 1922 when they met to consider Archibald Henderson's report on the necessary improvements to the Inner Harbour, there had been a sharp spat over the accuracy of the Minutes of the previous meeting. John M Morrison and Kenneth Mackenzie had moved and seconded them but Roderick Smith had made a counter motion that 'in view of the fact that the Minutes do not fully convey a true record of what happened at the Meeting held on 7th March and that the Finance Committee have over-stepped their remit, the Minutes be not adopted as read.' He was seconded by George Stewart but they received no support and the Minutes were adopted, Councillor Stewart dissenting. Archibald Munro recorded in the Minutes of this meeting that this vote was six to two.

However, at their monthly meeting on 4th April George Stewart queried that and maintained that he 'could swear before his Maker' that Councillor Maclennan had voted with them on 22nd March. Councillor Maclennan had told him that he had remained neutral and Councillor Stewart said he was quite willing to leave it at that, but even so this meant that the Minutes were not correct and the vote was five to two. John M. Morrison took issue with this, asking why Councillor Stewart should come there and dictate to them. Roderick Smith then weighed to say that Councillor Morrison should not be allowed to dictate to the meeting either. He had acknowledged that Councillor Maclennan could have been wrong and if he was then the Minute could not be right. What was required was a correct record and not what Councillor Morrison or Councillor Stewart thought.

The Chairman tried to move the meeting on by saying if there had been a mistake, it was his because he had counted the hands and given the numbers to the Clerk. George Stewart replied that Alexander Maclennan's word to him was that he had remained neutral. Captain Fletcher, who does not seem to have had any of Charles Orrock's patience and tact in these situations, snapped, 'I have nothing to do with what any member says outside this chamber .Councillor Morrison moved that the Minutes be adopted. Is that seconded?'

Ex-Provost Maclean said he would second. George Stewart held out, moving an amendment that the word 'six' be altered to 'five' and Provost Smith seconded, saying he did so 'in the interests of truth and of a fair record in our Minutes'.

This remark sparked off a furious reaction. John M. Morrison demanded that the Provost reconsider and withdraw the objectionable remark 'in the interests of truth'. The *Stornoway Gazette* reported that this 'led to an altercation across the table during which Councillor Morrison, ex-Provost Maclean and Councillor Bain rose for the purpose of leaving the meeting as a protest against time being wasted by Councillor Stewart and Provost Smith.'

Councillor Morrison and ex-Provost Maclean allowed themselves to be persuaded to stay because of the important business on the Agenda and sat down again. Louis Bain, however, said he was not going to put up with this sort of thing meeting after meeting and, according to the *Gazette*, put on his hat and left the building.

The flames were still smouldering round the table and Murdo Maclean again complained about the use of the term 'in the interests of truth'. He had been sitting opposite Councillor Maclennan at the previous meeting and his eyes were as perfect as the Provost Smith's any day. The Provost complained that there was no need to make personal remarks. 'Be a gentleman,' he said, 'and withdraw. I made no personal remarks.'

'You did,' accused ex-Provost Maclean.

'Not with regard to you,' replied the Provost. 'I have been twenty five years in public life and I never saw the like of this carry on before,' marvelled Murdo Maclean.

'What can you expect,' intervened George Stewart, 'when you vote that black is white?'

This brought Murdo Maclean and John Macritchie Morrison to their feet again and again the Chairman managed to persuade them to stay. Captain Fletcher then found himself on the receiving end of Roderick Smith's blunt tongue. 'If you, Mr Chairman, made a mistake in the counting, I know you are gentleman enough to withdraw and let the matter end,' he said.

The Chairman, obviously holding his temper with difficulty, said he would not withdraw what he saw with his own eyes. There were six hands held up for the majority and he would adhere to that in spite of anything that would be said at that meeting. They were quibbling about the Minutes and he was going to put it to the meeting. John Macritchie Morrison said he quite agreed with the Chairman and then, as if unable to stop himself, he said that he had now attended three or four meetings of the Commission and the same tactics had been employed at every one of them. He believed it had all to do with the Collector's salary.

It was now the turn of Provost Smith and George Stewart to be outraged and Councillor Morrison said that he would leave that subject but he would ask Provost Smith again to withdraw the phrase 'in the interests of truth' and leave it as 'in the interests of a fair record'.

The reply was that the Provost considered the phrase very desirable words to have in seconding Councillor Stewart's amendment and that 'a fair record' seemed to him equivalent to 'the truth'. This led to another heated exchange and Captain Fletcher, aware that things were out of control, said, 'We cannot go on like this and we must have the matter settled. I will take a show of hands for the amendment.'

If only it had been that simple. Councillor Stewart and Provost Smith raised their hands and then Ossian Robertson, who appears to have kept his head below the parapet throughout the battle, seemed to hesitate. When Captain Fletcher inquired if he was voting for the amendment, Mr Robertson confessed that he had not voted at the previous meeting. Councillor Morrison remarked that he thought Mr Robertson **had** voted at the meeting and the Chairman again asked for a show of hands for the amendment. This time Ossian Robertson definitely voted with George Stewart and Roderick Smith.

As expected the other three Commissioners present voted for the motion to adopt the Minute, which meant there was a tie.

'I suppose I have a vote?' inquired the Chairman and Archibald Munro confirmed this.

'How are you giving your casting vote?' asked Provost Smith.

'For the motion,' replied the Chairman. 'Then I make my protest,' stated the Provost, echoed by Councillor Stewart.

'And I want to protest,' continued the Provost, 'that Mr Morrison, after discovering that Mr Robertson did not vote at the last meeting, still voted for the motion'

This time the resulting argument across the table ended with John M Morrison leaving the meeting, saying he had other things to do and could better employ his time. Murdo Maclean followed and the Chairman called a halt.

'The thing is become intolerable and I adjourn this meeting,' he said. Provost Smith questioned why he was doing so. 'You are adjourning this meeting because three members have left in face of the fact that you still have a constituted quorum?'

Archibald Munro said the Chairman had the authority to adjourn the meeting and Captain Fletcher refused to continue, saying the importance of the business to be discussed required as full an attendance as possible.

Roderick Smith had the last word. 'For that purpose I am agreeable to your decision but I feel I have morally got the best of it,' he said.

The Collector's salary came up again after the two month period and the Commissioners agreed to promote William Grant to Assistant Collector at an additional £3 per month. They cut Angus Macleod's salary down to £7 per month, which was just short of a third of what it had been, and agreed to continue this for a further three months. Writing to ask Captain Fletcher to let him know if William Grant accepted the arrangement, Archibald Munro observed that 'I will have to write Mr Macleod, intimating the reduction of his allowance.'

When he did so, the Clerk expressed the hope that after three months the Collector would have improved enough to resume work again but it was not to be. Angus Macleod died on 27th August 1922 at his home at 14 Scotland Street. He was only 52 and left a widow and four children.

Angus Macleod was born in 1870 at 22 Cromwell Street, son of Allan Macleod, shoemaker and his wife Catherine. He had begun his working life at the age of 12 in the office of John Norrie Anderson. He had then gone on to serve an apprenticeship in marine engineering in a Glasgow shipyard, before applying for the post of Assistant Harbour Collector in his home town. As Roderick Smith had admitted, the Collector occasionally clashed with his employers and, judging from his remarks quoted in reports of the Town Council meetings, he was not intimidated by the fact that he worked for some of his colleagues in another sphere. He is described as having a kindly humour and a fund of anecdotes, with a cheery word and smile whenever one met him. Archibald Munro sent an Extract from the Minute to Mrs Macleod which recorded the Commissioners 'deep regret at his lamented death and also their sincere sympathy with his sorrowing widow and family in their bereavement.'

The Clerk also informed her that the £7 allowance would be sent to her on the 11th September but it had not yet been decided whether any further allowance would be paid.

William Grant was appointed as Harbour Collector at a salary of £220 per annum, having to provide security at £500 for the faithful execution of his duties and William J Macdermid was promoted to Assistant Clerk at 30s per week. At the same meeting the Clerk's annual salary was reduced by £10 to £120, while the Harbourmaster's went down by £20 to £225. Malcolm Stewart and Finlay Smith also found themselves with 2s per week less in their pay packet.

Leaving for a land of bright prospects

1923 was a momentous year of change for the island of Lewis. It was the year of the sailing of the first and most famous of the emigrant ships *Metagama* and the year when the people of the parish of Stornoway became proprietors of their own land. At the beginning of the year there was no indication of Lord Leverhulme's intention and the depressed state of the island's economy continued. On the other side of the world the growing nation of Canada was in need of manpower for its farms and industries and adverts began to appear in the Scottish press offering a new life for those willing to take the step. There were a number of schemes under which they could go to Canada and the one under which the *Metagama* emigrants left was the Ontario Scheme, financed jointly by the Ontario Provincial Government and the British Overseas Settlement Association. To be eligible for this scheme one had to be young, single and willing to work solely on the land, the men as farm labourers and the women on domestic service.

When the advertisements for the Ontario Scheme appeared in the *Stornoway Gazette* in January 1923, they received a huge response. Around 300 young people between the ages of 18 and 24 applied to the Ontario Government Representative at the shipping office of Murdo Maclean (not to be confused with the ex-Provost of that name) on Point Street. The Canadian authorities sent R S Percival, the Assistant to the Agent-General of Ontario, to the island to organise the departure and it was decided that a Canadian Pacific Liner should make a special call at Stornoway to allow the emigrants to embark directly. In early April the Commissioners considered a letter from Murdo Maclean, the shipping agent, inquiring whether any harbour dues would be chargeable when the SS *Metagama* arrived to ship passengers on the 23rd.

The Clerk wrote the Harbour Collector the following day to say the decision had been that 'if the SS *Metagama* comes within the Harbour limits she will be charged with Harbour Dues. The tender taking the luggage out to her is also liable.'

The tender was the *Hebrides* which arrived at the harbour early on Saturday 23rd April . Then around 10 am the *Metagama* herself appeared and dropped anchor off Stoneyfield. The quay was crowded with emigrants and relatives, officials and well-wishers, and the largest contingent of pressmen that the town had ever seen , including the new species known as 'kinematographers'. The necessary medical checks were carried out in one of the large sheds on the quay and when they had passed through this, the emigrants came out on the other side of the building to a cordoned off area leading to the gangway of the *Hebrides*.

Rev Roderick Morison of the United Free High Church presented each of them with a Gaelic Bible. A contingent of uniformed Girl Guides moved through the crowd, helping to carry the female passengers' hand luggage and generally acting as messengers. A huge bouquet of flowers from the gardens of Lews Castle went aboard the liner. Among its blooms were the little blue petals of forget-me-nots and the accompanying card sent sincerest, heartiest and best wishes from Viscount Leverhulme.

The *Hebrides* made two trips to the liner to transfer emigrants and on the first of these she also took out members of the Town Council and Harbour Commission to be entertained to lunch by the CPR company. Major Duffy, the company's representative, welcomed the Commissioners and Councillors, and Captain Fletcher replying on behalf of the Harbour Commission, said that he hoped the day was not far off when they could offer better facilities and when that day came they would look to Major Duffy to make Stornoway a regular port of call.

Bailie George Stewart replied on behalf of the Town Council in the absence of Provost Mackenzie, and began by saying that the pleasure of seeing such a greyhound of the Atlantic as the *Metagama* in Stornoway harbour was tinged with regret for them by the thought of such a number of their young men and women leaving their native shores. He spoke of the bitterness of parting but offered the thought that this voyage was entirely different to the voyages that these very same young men had embarked on since 1914. Then they had been destined for the blood-stained fields of Flanders or Gallipoli or some other horrible theatre of war. Today they were setting out, full of hope and confidence, in the peaceful pursuit of livelihood in a land of bright prospects.

Before they were taken ashore again the visitors were shown around the ship and pronounced themselves delighted with comfort that their people would be travelling in. Councillor Julia Fraser was remaining on board to travel with the emigrants and see them settled in the new country. Just after 5pm the *Hebrides* took the last group of emigrants out to the liner and this time she carried the Stornoway Pipe Band playing familiar tunes. At 6.30 she drew away from the the *Metagama* and a burst

Despite the smiling faces in this photograph, the sailing of the emigrant ships was another heartache for a community still mourning the loss of so many young men in the war and on the *Iolaire*

Malcolm Macdonald

The CPR liner *Metagama*, most famous of the three emigrant ships which took the youth of Lewis to a new life in Canada in 1923-24

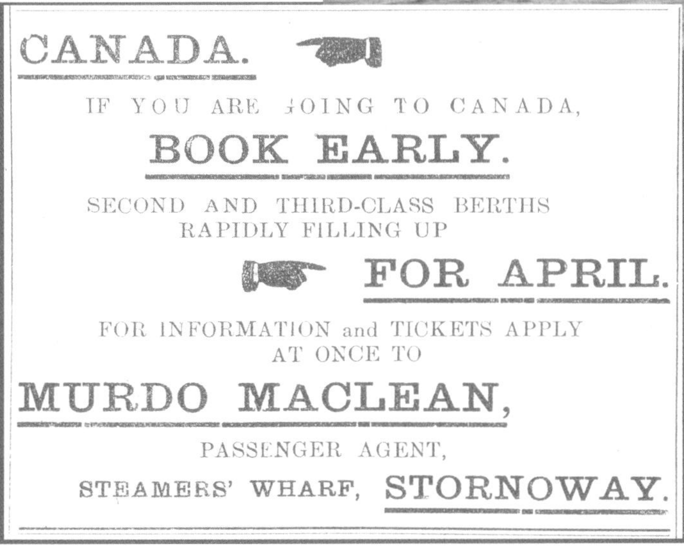

CANADA.

IF YOU ARE GOING TO CANADA,

BOOK EARLY.

SECOND AND THIRD-CLASS BERTHS RAPIDLY FILLING UP

FOR APRIL.

FOR INFORMATION and TICKETS APPLY AT ONCE TO

MURDO MACLEAN,

PASSENGER AGENT,

STEAMERS' WHARF, STORNOWAY.

Advert in the *Highland News* of 28th January 1911. At the turn of the 20th century the Canadian government actively sought immigrants from Britain to work on the prairie farms and on the railways of their fast developing nation

Norman Mackenzie

Relatives and friends crowd on to King Edward Wharf as the tender takes the departing emigrants out to the *Metagama*, 23rd April 1923

A year later sadness is etched on the faces of this group as they await the sailing of the *Marloch*

Norman Mackenzie

of cheering went up from both vessels. Almost immediately the liner took her pilot aboard, weighed anchor and steamed out of the harbour, watched by the crowd on the quayside. A bonfire burned on a headland near the harbour entrance and on the harbour flagstaff three flags hoisted by Captain Watt sent out the message: 'I wish you a safe voyage'.

In April 1923 the attractions of Australia were also being put to the Lewis folk by Mr Stillman of the Australian Government Emigration Department. They were looking for tradesmen and farm workers and in August the quays again rang to cheers and shouted good wishes when 100 islanders left for London to embark on the P&O liner *Bendigo*. Most were young single people but there were also a number of families who had been granted parcels of land in Western Australia.

The *Metagama* would be followed in 1924 by the *Marloch* and the *Canada*, each name synonymous for that generation of Lewis people with the hurt of parting. In a decade the island lost around 2000 of its young men to war, to the sea, and finally to the New World and it would take a long time to recover.

Provisional Plans

For the Harbour Commissioners 1923 had began with their dispute over the Council representation on the Harbour Commission. Having put that issue behind them they continued to work on the new Provisional Order throughout the year and Archibald Munro's letters to the solicitors Hume McGregor in Edinburgh take up most of the pages in the Letter Book. In February he wrote directly to Lord Leverhulme on the subject of his Lordship's harbour plan.

> 'Some considerable time ago your Lordship expressed a desire that if and when a New Provisional Order was to be applied for, that the Order should embrace your Lordship's proposed Outer Harbour Scheme as shown on a Plan prepared by Mr B P Wall dated 3rd February 1920. At a meeting of the Commissioners held last night I was instructed to ascertain from your Lordship if you still desire the Commission to apply for power to embrace your Lordship's proposed Outer Harbour Scheme. It is, however, understood that it is not now your Lordship's intention to proceed with your Scheme but the Commissioners would like to know definitely if this is so before the framing of the New Provisional Order is finally adjusted.'

Lord Leverhulme's reply was read at their meeting of 23rd March 1923. He said it was extremely difficult for himself to resume the development work on the scheme but he did not say that under no circumstances and at no time would he carry it out. He was compelled to say at present that he did not see any immediate possibility of resuming the work and therefore the Commission would be able to decide for themselves whether the New Provisional Order should embrace the Scheme and he must leave the decision definitely in their hands. The meeting that day was in the hands of Provost Kenneth Mackenzie and the decision was delayed until Captain Fletcher was present, when it was agreed to find out from Hume McGregor what plans and other details they would need for the Provisional Order and any other guidance the lawyers could give. At their Annual Statutory Meeting they took the decision to incorporate the Outer Harbour Scheme in the Order and asked Hume McGregor to have a long period for executing the work incorporated as well.

Hume McGregor had an informal discussion with Mr Watson KC, who had already advised them on their borrowing powers and reported their conclusions to the Commission in June. They explained that there were two ways to proceed with a new Provisional Order. One was under the Private Legislature procedure which would allow a longer period of time for repaying loans, sometimes as much as 60 years. The other way was through the Piers and Harbours Act which would mean it would proceed under the auspices of the Board of Trade. The period of repayment in that case would be very much shorter. If the Commissioners intended to include the whole of the works proposed in the 1920 plans, then they should be aware that all Orders had a time limit imposed within which the works would have to be commenced, usually 3 to 5 years but sometimes 7 years. If Lord Leverhulme did not intend to commence this scheme within the next three years or so, it might be better to frame the Order for the powers required by Commissioners themselves, so that the borrowing powers and the Inner Harbour improvements were not affected. The Commissioners deferred consideration of this advice until their next meeting.

The Gift

At this stage, although there is no reference to it in the Harbour Minute until weeks later, the sensational news had broken of Lord Leverhulme's decision to leave the island and offer it to its people as a free gift. His leaving was a shock in itself but the offer, which *The Scotsman* reckoned to be close on £1m in value, was astounding.

As Councillor Alexander Maclennan said at the time, it was one of the most extraordinary offers that had ever been made to any people in Scotland. A landlord giving land free to his people was unheard of and it threw the island into a fever. Lord Leverhulme proposed to hand over his interests to two separate Trusts, one to include all the land within a seven mile radius of Stornoway, the other to administer all the lands outside that boundary. He gave them barely a month to consider the offer. The Councillors and ratepayers of Stornoway accepted almost immediately but the Lewis District Committee, after much heart-searching, voted against.

Although the Harbour Commission as a body was not involved in the decision, the outcome would affect it considerably. Writing to Hume McGregor on 27th September 1923, Archibald Munro told them:

> 'the pushing forward of the Provisional Order is delayed meantime until a decision is come to by the Local Authorities with regard to Lord Leverhulme's Offered Gift of the Lewis to them, as, if the Gift will be accepted, a modification of the present Constitution of the Harbour may have to be provided in the New Order.'

The first time the subject appears in the Minutes is at the 11th December meeting when a letter was submitted from Lord Leverhulme's Edinburgh agents intimating his Lordship's free gift of a number of heritable properties around the harbour area. The annual rental of these was shown as £1,327. On the motion of Provost Kenneth Mackenzie seconded by John Macritchie Morrison the ten Commissioners unanimously agreed to accept the gift with the deepest gratitude and:

> 'they record their high appreciation of Viscount Leverhulme's munificent gift to them for which they tender him their profound thanks but regret very much that he seen it to be his duty to sever his connection with the town and island.'

With that acceptance the Commissioners became landlords of a variety of buildings from North Beach to Newton, including some on Point Street, all in various states of repair. Munificent though it was, it still presented considerable administrative problems and expense. There was insurance to be arranged, rents to be collected, arrears to be pursued. Was No 12 Point Street suitable as a dwelling house? (to which the answer was no). Repairs were needed to the drains and water supply at No 19 Newton. What would be the cost of removing the range in No 9 South Beach and fitting in the tenant's new stove?

For Archibald Munro this was only part of the correspondence crossing his desk in the Masonic Hall. His main employment was Town Clerk and as such he was deeply involved in all the changes affecting the burgh that autumn. At the same time, as Clerk to the Harbour Commissioners, he was dealing with all the financial and legal aspects of the proposed improvement works. In September he received a letter from the Harbourmaster Captain George Watt tendering his resignation, given as his reason that 'in his declining years he finds that the long distance between his home and the harbour is too harassing upon him especially during the winter months' and in October the Public Works Loan Board wrote to warn that if the outstanding repayments were not made within ten days, they would appoint someone to collect the sums due.

The PWLB letter was submitted to the 30th October meeting of the Commissioners who decided to let it lie on the table, but within days another letter arrived from Colin Scott Mackenzie, the Stornoway solicitor appointed by the PWLB, calling attention to the arrears and asking if they had any proposals to make. The Clerk was instructed to ask for more time 'as up till last year they have met all their obligations to the Board, and they do not want now to be relieved of them, but that time be given them until the principal trade of the port improves, as there is every prospect that it will, when all the sums due to the Board will be paid.' This seems to have been accepted and the money was sent on three months later.

The Harbourmaster's resignation took up more of the Clerk's time. His advertisement for a successor brought in 88 applications which a Sub-committee whittled this down six. The full Commission took this down to three for interview – W R Mackintosh, Kirkwall; W J Nicolson, Leith; and Alexander Finlayson whose address is given as Glasgow. Captain Nicolson wired his withdrawal, saying the £225 salary was too small, and Captain Mackintosh was en route to New York and would not be available until mid-January. Captain Finlayson, however, was interviewed and then had to wait another week until the Commissioners came to their decision. This he received in a letter dated 27 January 1923 appointing him to the post as from 8th January 1924 'with all the rights and privileges conferred upon a Harbourmaster by the Stornoway Harbour Orders 1865, 1881 and 1892 and other Acts of Parliament'. One condition of his employment was that he should reside within the burgh. Another was that he should supervise all ordinary work in connection with the harbour property.

On 15th January 1924 the Minutes show that 'the newly appointed Harbourmaster' was present as they discussed the Sgeir Mhor Perch, the gifted properties and the Provisional Order. Alexander Finlayson remained with the Commission for almost 40 years, becoming a legendary figure in the town's history. He had begun his working life as a pupil-teacher in his home village of Bragar but left as a young man to join the Merchant Navy. Now just over 40 years of age he had worked his way up from Able Seaman to Captain and during the recent war had been a Lieutenant Commander in the RNR where he had served in a flotilla of the Dover Patrol. Since then he worked for the Burns Laird company, running between Glasgow and Dublin.

‘Four and a halfs’

Since his arrival in 1924 the Harbourmaster had been waging war on the practice of supplying and drinking beer on the quays. This was a consequence of the prohibition on the sale of alcohol which had been in existence in the town since 1920, after a vigorous temperance campaign. Some members of the community had found a way to circumvent the rules, which was to retire to a corner of the harbour with a 4 $^1/_2$ gallon container of beer and share it with their companions. As long as no money was seen to change hands, the authorities could do nothing to stop the practice.

In March 1925 Captain Finlayson reported that he had put up notices prohibiting ‘loitering and trafficking in alcoholic liquors’ on the quays and he had ‘personally requested groups of men found contravening the order to remove which they declined to do.’ He had reported the matter to the Superintendent of Police and asked for his support which he had not received. Archibald Munro was then enlisted to back him up and wrote to Inspector Campbell ‘with regard to the annoyance caused by the congregating of numbers of people on the piers consuming beer and to ask for your assistance in trying to put a stop to this reprehensible conduct.’

Inspector Campbell replied that as far he could judge, the gathering of people on the piers for consuming beer did not constitute an offence. No case of wilful obstruction had come to the notice of the police and if it did they would take action.

The Harbourmaster reported back to the Commissioners that he had only been partially successful in carrying out their instructions. The present area used by the beer drinkers was confined to the vicinity of the Mortuary ‘where they are a nuisance and an obstruction’ but having reported a case of refusal to remove to the Police, and no action having been taken, he was awaiting further instructions.

Archibald Munro wrote to the Procurator Fiscal this time. ‘I have been directed. . . to inquire whether it is intended to bring the case before the Court, and, if not, that you give the reason for not doing so.’

Colin Scott Mackenzie replied that he did not consider there was sufficient evidence on the one case reported to him.

Then the fishcuring company of George Cormack and Sons complained that their curing plot was being used as a urinal by the beer drinkers. Herrings and barrels were being destroyed, they said, and the smell was injurious to the health of their workers. The letter was handed to the Procurator Fiscal and the Commissioners agreed to interdict those congregating on the piers to drink. A notice was issued to the press giving 7 days warning of the intention. In November 1925 Captain Finlayson submitted three names of persons who 'persisted in serving beer and who refused to move when requested to do so.'

Finally in August 1926 the Clerk of Court was able to report that the case had been decided but he had not yet received the details of the Sheriff's Interlocutor. When this document was received it was remitted to a Sub-Committee for consideration and it remains in a slim file in the harbour archive. It shows that the defenders maintained that they had received no intimation of the decree against them and that, in any case, the interdict was invalid because it banned them from using certain places for purposes which would have been perfectly lawful on a public highway.

The Sheriff-substitute replied: 'The defenders admit on Record that they consume beer on Cromwell Street Quay but there is no clear indication or admission on record that the place on that Quay where they do so is within the premises of the pursuers as described in the Decree of Interdict.' He gave both parties a week to sort out the ambiguities in their pleadings, at the end of which he dismissed the case against Thomas Brown, but fined Murdo Macleod, Murray Beaton, Norman Mackenzie and George Macleod, £5 each.

The Commission's Sub-Committee recommended that 'in view of the expenses involved and the unsatisfactory results from this form of procedure for obtaining the end in view, that some other form of procedure should be adopted, if possible, to put a stop to what is considered to be an outrage on public decency.'

About a week later the general public solved the problem by voting the town 'wet' again by 477 votes to 393. ❏

The early part of 1924 shows him setting to work with a will, reporting on the repairs needed for the wooden wharves, the necessity of painting all the beacons in the harbour, and the state of the road between the curing stations and South Beach retaining wall between No 2 Pier and the Patent Slip. Portions of this stretch were so bad, he reported, that horses had great difficulty in hauling their loads and on the curing stations themselves there were so many hollows that the water could not run down into the gutterways. He suggested gravel from the Laxdale river and reckoned £100 would put the roads 'including the levelling of hollows and cart ruts around all the quays' in good condition. In order to preserve the harbour property he suggested painting the Fish Mart roof and all the corrugated iron sheds belonging to them. He thought the Flag Pole on the site of the old Castle at No 1 Pier looked in need of attention and found that it was so rotten it might have fallen down at any time. The greenheart 'tabernacle', however, was in good condition and if the Commissioners wanted to replace the pole he would suggest one of 55 feet instead of 60 feet. He considered the method of lighting the harbour 'wasteful and out-of-date' and recommended that, to save mantles and glass, all the lanterns at present fitted with upright mantles should have downward ones fitted and a door to be fitted in all mantles to allow them to be lit with an acetylene torch.

One of the new Harbourmaster's earliest requests involved the gift of the Leverhulme properties when he asked if there was any possibility of being allocated a house on South Beach. The Commissioners said they would be only too pleased to accommodate him when a suitable house came into their possession. The upkeep and administration of these gifted properties was time consuming, and by the summer of 1924 they had taken a decision to appoint a Factor to look after them at £80 per annum. Eight applications were received, including William Grant, Harbour Collector, W J Clarke, Sheriff-Clerk Depute, and C Scott Mackenzie, Solicitor. Edwin Aldred, 41 Matheson Road secured the post, beating the Collector by one vote. 'It will be necessary for you to find security for £800 for the faithful execution of the duties of the office,' the Clerk advised him. 'The Commissioners will pay one half of the Guarantee Fidelity Premium.'

Edwin Aldred was given a separate bank account with £100 to start him off and authorisation to carry out repairs costing up to £15. Anything estimated over that would have to go to the lowest tenderer. The insurance of the properties was halved between Caledonian Insurance Company and the Royal Insurance Company 'on the understanding that the local agents for both companies get the Commission.' Robert Thompson, Clerk of Works to the new Stornoway Trust, was retained temporarily to report on any work requiring to be done. The Titles to the Properties were received from Lord

Leverhulme's lawyers and they remain in the Harbour Office archives today.

At 2.30 pm on 5th September 1924 the Harbour Commissioners met as a body on the hill above Stornoway to see Lord Leverhulme unveil the Lewis War Memorial, on his last visit to Stornoway. Captain Fletcher was present, but was no longer their Chairman. His duties as Chamberlain had ended a few days earlier and he was no longer eligible to sit on the Commission. The Commissioners recorded their deep regret at losing his valuable services and wished him every success and happiness in the new sphere to which he had been appointed. Archibald Munro, sending the Extract of Minute on to Captain Fletcher at 'The Cottage' on Goathill Road, added 'I need hardly say that I associate myself very cordially with the resolution and have to thank you for the very kindly reference you made to myself in your letter.'

Modestly he has not recorded that reference in the Minute.

The new Stornoway Trustees were now obliged to appoint a Chamberlain who could act as Chairman of the Harbour Commission under the existing legislation. They submitted an extract from their Minutes appointing Mr Hugh Macleod to be the Estate Clerk and Factor. The Harbour Commissioners accepted this as 'sufficient evidence of Mr Macleod's appointment as the Superior's Chamberlain' and accordingly Hugh MacLeod too k the chair in November 1924.

Changing Orders Again

1926 was the year of the General Strike in Britain, the year when the work of rebuilding the Town Hall began, and the year when the new Harbour Order finally came into force. It had taken over a year to get it from Draft to Royal Assent.

In October 1924 Archibald Munro had written to Hume McGregor:

> 'I was patiently awaiting a wire from you intimating that you had forwarded the Draft Order but I received this morning your letter explaining the reason for the delay. . .'

This seems to have been caused by the acquisition of the Gifted Properties because the Clerk ends his letter glumly:

> 'I expressed to Capt Fletcher on more than one occasion my doubts as to whether it was competent for the Commissioners to have accepted of the properties gifted.'

However, the draft had appeared very shortly after this exchange and over the last few weeks of 1924 the Commissioners had met regularly to go through the Draft Provisional Order and make alterations and additions.

Before finally submitting it, they asked Hume McGregor to find out if a further sum over the £60,000 could be included. Stornoway Trust confirmed that, as Superiors, they would guarantee a feu duty of £6 per annum for such part of the foreshore as the Commissioners might need for the Harbour Extension. The Commissioners had considered the reconstruction of No 1 Wharf by means of sheet piling or reinforced concrete but the estimate of £20,000 put them off the idea, and they withdrew the proposal to include a sum for that in the Order.

They instructed Messrs Robertson and Carphin, Chartered Accountants, Edinburgh to prepare the statement needed to apply for fresh borrowing powers. Robertson and Carphin had been appointed the Commission's Auditors the previous year and the firm would remain the Commission accountants almost to the end of the century. The Trust Clerk of Works made a tracing of the Estate Plan and plotted out all the new Harbour properties on it for forwarding to the Engineers.

In April 1925 the draft order was issued, stating what the Harbour Commissioners intended. The three previous Harbour Orders were to be repealed and all their provisions consolidated in the new one. The Constitution was to be altered and the qualifications for the Commission's ten members were to be redefined, as were their powers and the limits of the harbour undertaking A new wharf was to be constructed.

Then just when it seemed they were ready, MacBrayne's threw a spanner in the works by opposing the Order. The Company said it was doing this 'in no militant spirit but to guard our interests'. They wished to know what rates would be chargeable on the mail steamer and suggested a return to the pre-war stystem of compounded dues.

Replying to the Commissioners' request for their views, MacBraynes set these out in a letter in May 1925. They wanted regular calling vessels such as the mail boat to be charged on the tonnage of traffic carried rather than on that of the vessels – 'thereby accepting the basis of trade as against frequency of calls, which are for local convenience'. This proviso, they said, had been long recognised as a reasonable basis, and had been incorporated in all the recent Harbour Orders in the West Highlands. It was unanimously agreed on the motion of Roderick Smith seconded by Norman Maciver, that MacBrayne's be told that the Commission was not prepared to accept the proviso but in the special circumstances they were prepared to give consideration to a compounded sum of dues on the mail steamer.

Mr McGregor of Hume and McGregor came to Stornoway to go through the evidence to be presented in support of the Provisional Order which went to the Parliamentary Committee in Edinburgh at the end of July 1925. Those nominated to appear were the Chairman, the Clerk, Provost Mackenzie, Bailie Bain, John Macritchie Morrison, and George Stewart, as well as Mr Hall of Messrs Leslie and Sons, Aberdeen and Captain MacCallum of the SS *Renaldo*. It does not seem as if either Mr Hall or Captain MacCallum attended but all the others with the exception of Bailie Bain made the journey, as did Robert Thomson, the Clerk of Works. The Clerk wired Hume McGregor on 25th July:

> 'Stewart can speak to freight goods mail boat other than herrings also Macritchie Morrison stop – Stewart Morrison and myself arrive Monday night – Thompson left last night but will be present at enquiry Tuesday forenoon – Bain not going – Morrison puts up at County Hotel where his wife is at present.

At this point Archibald Munro was around 77 years old and still employed as Town Clerk as well as Secretary/Treasurer to the Harbour Commission. Earlier in the year he had replied to Mr McGregor 'I am obliged for your inquiry as to my health. I am keeping in a fairly good condition, all things considered.' As was said of him later, he seemed to have the secret of perpetual youth.

Archibald Munro

Stornoway Historical Society

Julia Fraser

Fraser family

More Delays – the 1930s

The 1930s saw the next major stage in the improvement of the harbour facilities, long delayed though it had been As the decade opened, it seemed that the Commissioners would able to fulfil at least part of the promises they had made to the fishermen, one of which they had made to the Buckie Fishermen's Association in early 1929, pledging they would 'do all in their power to give facilities for accommodating herring drifters and the discharging of herring at the quays.' Finance had been

A Woman on the Commisision

The first woman appointed to the Harbour Commission arrived in May 1932 in the shape of Councillor Julia Fraser. Mrs Fraser, by then sixty years old, had begun her political career over a decade earlier, standing for election in 1919 as soon as women received the vote. Born Julia Martin at Crobeg Farm in Lochs, she worked in America as a nurse before returning to the island and marrying William Fraser. She served on both the Ross-shire County Council and the Stornoway Town Council and was the town's first woman Bailie and magistrate of the Burgh at a time and in an island which later generations sometimes depict as too reactionary and chauvinistic to allow such a thing.

Sharp and intelligent, Julia Fraser spoke her mind without fear of the consequences and was frequently at bitter verbal odds with other Councillors. At the Ratepayers' raucous public meeting before the 1931 election she had welcomed another lady candidate, Mrs Isabella Whitaker, to the hustings. 'I have heard that the slogan for the election is "Keep out the ladies; keep out Mrs Fraser whatever happens to the rest",' she told the lively audience, 'but I want to say that they cannot keep women out of anything at this time of day.' The voters did not keep them out.

'Mrs Whitaker, we are going to put you in,' shouted a heckler from the Town Hall floor in November 1931. Both forthright, fearless, women were elected and their individual contributions to public life went down in local history. Mrs Whitaker did not sit on the Harbour Commission but half a century later her grandson would be appointed its Collector of Dues and later its General Manager. ❑

secured, the Scottish Fishery Board were happy, all the plans were in place, and the work ready to start.

But the New Year celebrations were barely over when everything changed yet again. The Public Works Loan Board now had before them the second application for an extra £6,000 and they wrote in concern. 'This Board do not understand,' their letter said, 'why it has become necessary to dispense with the underpinning works at North Beach Quay. . . and to alter the whole scope of the permanent works by constructing a Reinforced Concrete Wharf along the whole length of that quay at a cost of £15,000 in place of the Reinforced Concrete Breastwork authorised by the Order of 1926 and estimated to cost only £8,900. . .'

The Board felt that the cost of the permanent works was being increased at the expense of the necessary dredging and underpinning work. The Consulting Engineers, Henderson and Nicol, produced explanatory statements and drawings, but to no avail. The Public Works Loan Board were not convinced and in March 1930 they withdrew their offer of loan. They explained that since the loan had been granted, the Harbour plans had 'undergone a radical change which will necessitate the case being considered *do novo*'.

A fresh application must be made through the Ministry of Transport.

The Commissioners delayed a reply to this letter in the meantime and went ahead with advertising for tenders for the construction of the reinforced concrete works.

By this time the Commission had a new Chairman, Norman Maciver, of the Stornoway kippering and coal firm Duncan Maciver Ltd. At 42 he was, said his proposer, John Macritchie Morrison, 'a young man possessed of outstanding business ability and with a considerable interest at stake in the community.' Norman Maciver's appointment was unanimous and he has been credited by James Shaw Grant, then the young editor of the *Stornoway Gazette*, as being the driving force which got the harbour improvements through the corridors of Government power and the miles of red tape in the way. The Minutes of the time certainly show that much of the early correspondence with the various Government departments was drafted by him, although revised and approved by the Commission. In a tribute to him after his death in 1960, two of the phrases used to describe him were 'forthright of opinion' and 'blunt of answer', neither characteristics calculated to make him popular. This did not worry Mr Maciver. The writer states that he 'was less concerned with popularity than with getting things done.'

Norman Maciver and his fellow Commissioners were determined to get the improvements to Stornoway Harbour done. Another of his aims in 1930 was to curb the high cost of collecting harbour dues and to that end he moved that the two main cargo operators, Coast Lines Ltd and Messrs David MacBrayne (1928) Ltd, should be approached to see if, for a 5% commission fee, they would collect the dues on inward and outward goods carried on their vessels. He also moved that the regular Herring Trade Shipbrokers operating at the port should be asked to do the same. He added that since these shipbrokers already had to collect all dues at all the principal herring ports in Scotland and England, without commission, they would be quite prepared to do the same at Stornoway.

MacBrayne's in the 1930s, as its successor Caledonian MacBrayne does today, carried a host of different commodities addressed to individuals all over the island. Nowadays these arrive in the hauliers' vans for distribution. In the 1930s they were collected from MacBrayne's store by carriers from various parts of the island who naturally would not pay out the dues on behalf of the consignee. Holding on to the goods until the consignee was contacted and paid up was going to take up storage space as well as add to their work load. The dues payable on these small consignments were also very small, amounting to pennies.

In August 1930 the Harbourmaster received another complaint from MacBrayne's, this time because one Friday night that month the local manager had to stop the discharge of the mail boat's cargo owing to the 'impossibility of the getting trucks through the crowds assembled on the pier.' At this time Stornoway Harbour was the only gateway to the mainland for travellers, with air services still a number of years in the future. Watching the arrival of the steamer had been a daily ritual for Stornowegians for years. In the words of a *Gazette* columnist of the day, J N Maciver, 'In our town at this time of the year the daily tonic of seeing the mail steamer come in is much greater and more effective than at any other time.'

Finlayson family

Captain Finlayson

Jack Maciver's article, which had appeared in June, was entitled 'The Gaoler' and bemoaned the Harbour Commission's efforts to control access to the piers. He described the evening ritual as a harbour employee, 'a man with brass buttons', went about the business of closing off the piers to the general public. The 'gaoler' started at the gates near the Harbour Office 'and magnificent works of art and invention they are, all steel and complexity and they slide into their place in a really marvellous fashion.' The gaoler then moved on to more gates at the Fish Mart and finally roped off the other entrance to the wharf at MacBrayne's store. The result of all this, said the writer, was a long queue extending past the Fish Mart who were determined to get as near to the boat as they could. A good number of that queue, he said, could recount 'all the details of the steamer's arrival, who came that he or she knew, and all the interesting stories of that glad event.'

Captain Finlayson's report on the matter is not detailed in the Minute but appears to have been concerned more with the departure of the vessel, which also attracted many people seeing off friends and relatives, as well as those with goods to despatch, not to mention the plain curious. After consideration the Commissioners decided to ask the Inspector of Police for the services of a Constable for about half an hour before departure. They also suggested that MacBrayne's should post two of the ship's crew at the entrance to the gangway to direct passengers and assist with controlling the crowds.

In August 1932 the Secretary reported he had applied to the Board of Trade for their approval of the Cromwell Street works. In September the Public Works Loan Board issued approval of a loan of £18,000 over 30 years at $5^1/_2$%. Consulting Engineer Archibald Henderson was summoned to Stornoway immediately to set out the next part of the programme. His recommendation that the work be carried out by direct labour was approved and he was also asked to make immediate arrangements to settle the price of the required materials, it having come to the Commissioners' ears that the price of steel might soon go up.

Donald Smith, Resident Engineer, reported that if the work was to be finished in time for the summer fishing season, it would be necessary to work one and a half shifts per day. He would therefore need moveable electrical arc lights in preference to acetylene flares. He was allowed to confer with the Manager of the Electrical Supply Company to see what could be set up. He was also allowed to install electricity in his own office and granted a travelling allowance of 17/6 per week for using his car in connection with the works.

In October 1933 the Commissioners finally received the Fishery Board's Certificate of Approval for the Cromwell Street Quay Works. The Consulting Engineers reported that the total expenditure on this part of the improvement scheme had been £19,438 12s 9d. Three months later, in January 1934, the Commissioners received the payout from the Public Works Loan Board of the promised £18,000 loan.

Some Light on the Subjects

The question of lighting the harbour was the cause of much discussion during the years of the improvement works. Until 1932 Stornoway was lit by gas, supplied from the Gas Works on Bells Road which had been part of Lord Leverhulme's gift to the Stornoway Trust. The Town Council annually appointed five members to the Trust, by virtue of which they became directors of the Gas Company. The gas supply had a tendency to be erratic and some buildings, such as Lews Castle and the Picture House, had their own electric generators by the late 1920s.

The new Town Hall was one of those and in November 1930, the Town Council deliberations in its brightly lit Convening Room were interrupted by a deputation of eight irate business men, drawing the Councillors' attention that the rest of the town was in darkness due to a failure of the gas supply. Furthermore, the *Gazette* report of the incident stated, the mail steamer was at anchor in the harbour unable to berth because of no guiding lights on the pier. This was the third time in recent months that the supply had failed and the citizens were becoming annoyed. Provost Maclennan repaired to the Gas Works to see what the trouble was and returned to assure the deputation that the directors of the Gas Board 'would take drastic measures to remedy matters'.

Several members of the community were aware that electricity was here to stay and decided the way ahead for the town was to have a new generating station and a proper electricity supply. To that end the Stornoway Electric Supply Company was formed in April 1932, with Norman Maciver as Chairman, and was up and running in record time. The new generating station was opened in September 1932, while the Harbour Commissioners were involved in the second part of their harbour scheme.

The lighting of the harbour area had became a problem as early as 1931 when it was discovered that the gas pipes under No 2 and 3 Piers were perished beyond repair. The leakage of gas was so bad that No 3 had to be cut off completely and repairs set in motion immediately. Captain Finlayson, Mr Arnott, the Manager of the Gas Plant, and Clerk of Works Thomson were instructed to look into the system of lighting the whole harbour but the result of their investigations is not reported in the Minutes. Perhaps there were too many other matters claiming their attention or perhaps they were delaying handling a political hot potato.

The next time the subject appears in the Harbour record is June 1932 when a letter was submitted from the Stornoway Electric Supply Company Ltd asking permission to install a number of Distribution Pillar Boxes on the walls at South Beach, North Beach and Cromwell Street Quays. The company also reminded the Commissioners of a letter that body had sent to Anderson Macarthur & Co, in November 1929 promising that if a company was floated 'for the purpose of supplying the town with electricity that they would give support as consumers provided the terms of supply were reasonable.' The Commissioners were cautious and replied that, without committing themselves, they would be willing to consider an offer for wiring and lighting the harbour. Absent from the meeting was Norman Maciver, who happened to be Chairman of the Electric Supply Company. Those present were William Macnab, Robert Steven, Louis Bain, William John Tolmie, Julia Fraser and the Chairman, Roderick Smith, who also happened to be a director of the Stornoway Gas Company. Some of the other Commissioners were also on the board of the Gas Company but do not appear to have drawn attention to this in the subsequent deliberations.

Finally the outcome of discussions was minuted that 'after very careful consideration' the Commissioners resolved to light electrically the west side of No 1 Wharf, Esplanade Quay, the Harbour Office, the Fish Market and the new store to be built on Esplanade Quay; also that all tenants of harbour property be given the option of introducing electricity to their respective subjects, that Henderson and Nicol be asked to report on the matter, and that the Stornoway Gas Company be invited to report as to the improvement of the existing gas lights at the harbour. Mr Arnott, Manager of the Gas Company, took up the invitation almost immediately, arranged a 'regular and systematic inspection of the lighting free of charge' and said he would afford the Commissioners any personal help or advice that may be necessary.

'The very centre and hub'

As the Cromwell Street works finished the Commissioners moved on directly to the next stage of the scheme namely the reconstruction of No 1 Wharf. In May 1933 the new Clerk Alasdair Macleod submitted a copy of an application to the Public Works Loan Board for a further loan of £23,000, along with the Loan Board's reply that they would consider it when the Ministry of Transport approval had been received. He had already applied for this and now had two copies of the list of queries for harbour authorities to answer when applying to borrow. Norman Maciver explained to the meeting that these steps had been taken

'Some of the Worst Hovels in Stornoway'

The state of the housing owned by the Harbour Commission was brought under the public spotlight in Tuesday, 19th March 1935 when a fire completely destroyed their tenement property at 18 Point Street. Six out of the seven families which rented accommodation in the building were made homeless and housed temporarily in Carn House, next to the Town Hall. The fire started in the attic of one of the houses in the tenement, possibly by a spark from a log fire in the room. It quickly took a hold in the roof and the building rapidly filled with smoke. The Fire Brigade were quickly on the scene, since all the Burgh workmen had been gathered in the stables when the call came in. Until they arrived in, according to *The Scotsman*, 'the modern fire engine gifted to the town some years ago by Lewis natives now resident in Chicago on the occasion of the re-opening of the new Town Hall' the police and neighbours tried to control the fire with buckets of water. The tenants on the ground floor were able to remove their furniture from harm. Those on the top floor, however, lost almost everything but the clothes they wore. The attic and first floor were completely destroyed by the flames, while the lower floor was badly damaged by the water from the Brigade hoses. The Gazette noted that the incident could have been much more serious. 'In the crowded area of Point Street where it occurred the flames might easily have passed from house to house over a wide area. The seven families affected comprise about thirty individuals, including children.' 'Why,' asked Councillor Macleod at the Town Council meeting in November 1936, 'should houses owned by individuals be condemned wholesale, and houses owned by one of our public bodies not be condemned? There are no worse hovels in Stornoway than those in Point Street!' ❑

because 'it was futile to apply for grant since they had been turned down so often.' By June 1933 the Secretary was able to report that the Ministry approval had been received and Roderick Smith commented that this was very satisfactory progress, since they had formerly had to wait a very long time for any advance.

However, perhaps the recent national herring dispute had repercussions in London since in August, although the Secretary reported that approval had been received from the Fishery Board for the plans and specifications for No 1, he also had to report a depressing response from the Public Works Loan Board. The PWLB considered that the income of the Harbour Undertaking did not provide sufficient security for the further loan applied for. The Secretary, along with the Chairman and Vice-Chairman, had drawn up a lengthy reply asking the Board to reconsider. They laid out facts and figures demonstrating an increase in revenue and the ability of the Commission to meet the loan repayments; they stated that 'the steadily growing commercial importance of the harbour and the pre-eminent position which the port held in the herring fishing industry' meant that revenue was unlikely to decrease; and they said that the improvement works meant less maintenance costs which might further boost the surplus. Their main concern was that No 1 Pier was now in such bad repair and in need of such expensive maintenance that it might soon have to be abandoned.

'After having accomplished so much within the past three years it would be a great pity if its hub should now be left in a dilapidated and unserviceable condition,' they said, pointing out that the wharf provided 'the only berth which could be approached at all states of the tide by vessels of a very high tonnage which were in part engaged in the export of cured herrings to foreign ports, and which formed a very lucrative source of revenue.'

Norman Maciver said that the Secretary's letter was a masterpiece and he did not see how the Loan Board could refuse. Ex-Provost Smith agreed but warned that the Loan Board was very much 'in the hands of their Secretary' who seemed 'more prone to refuse than to grant.' Roderick Smith's instinct proved correct.

The Commissioners decided, on the motion of A J Mackenzie seconded by William Macnab, to approach the Board once more and if that were unsuccessful, to appeal to the Treasury for a grant to complete the works. He moved that they write to Board again. He did not think it would prejudice their case to send the Treasury copies of all the correspondence, in case they had later to apply to them for grant.

This elicited a slightly more encouraging response to the effect that the Loan Board were prepared to accept early payment in October of the £125 principal remaining on the old loan. Alasdair Macleod's answers obviously settled the questions since at the beginning of January 1934, he was able to report the granting of a £23,000 loan at 4 $^{1}/_{8}$% over 30 years.

Alasdair Macleod was 25 years old and had taken over the duties of Clerk to the Commissioners on the death of Archibald Munro in March 1931. He served two years as Interim Secretary before the Commissioners formalised the appointment. The Burgh Commissioners were not so slow to make him Town Clerk. Their faith in him was justified by his 37 years service until his early death in 1968.

The *Stornoway Gazette* welcomed the news of the loan by saying that 'not only does it mean that the reconstruction of the harbour will be carried out in full as the Commission visualised it when they launched their ambitious scheme some years ago, but it will also provide work in the town at a time when it is much needed because of the failure of the winter fishing and the general trade depression.'

No 1 Pier was described by J N Maciver in one of his articles as 'the very centre and hub of all the town's activities.' He described it as alive with traffic and busy people, the favourite berthing place for herring drifters to be near the Market, and the site of much commotion when the steamers belonging to MacBrayne's, Coastlines, and other companies, were arriving and departing. However, he said, 'Number One is old and getting much the worse of the wear. It has an old-fashioned look about it and belongs more to the past than the present. It belongs to the days of horses and horse-drawn vehicles rather than to the days of the ubiquitous motor car.'

The Harbour Commissioners were set to change that. In March 1934 they accepted a £2,160 tender from Messrs F A Macdonald & Partners, Glasgow for the steelwork of the new construction, while the main construction contract went to H. M.Murray, Glasgow, at a keen price of £17,714 4s 3d. The £23,000 loan for the work on No 1 Wharf was received before the end of the financial year on 11th November 1935.

There seemed to be no end to the determination of the Commission to reconstruct the harbour wharfage. Work had barely begun on No 1 Pier when the subject of concreting No 2 was raised, at the same Special Meeting on 11 May 1934 as the idea of the new offices was first mooted. The Special Committee heard from its Convener, Norman Maciver, that he had been discussing the matter of office and storage accommodation with the Resident Engineer who had come up with a suggestion to reconstruct the existing building with an additional floor for office accommodation. This would entail widening the existing building and reconstructing the two end sheds under one roof, eliminating the entrance to the wharf between them. A second floor would be added for office space, and the removal of the lean-to building in which the offices were presently housed, would widen the entrance to the wharf opposite the Fish Mart and also the alleyway between the Mart and the Harbour Office. The west side of the building would run parallel with that of the reconstructed and quay and, if the Commissioners so desired, 'an ornamental design could be prepared for the southern face of the building' . The Committee approved the plans and instructed the Engineers to prepare the specifications and invite tenders for the work. The *Gazette* sub-heading on its report of the meeting proclaimed: 'A Handsome New Building for No 1 Wharf' and quoted Norman Maciver's remark that it would improve the amenities not only of the pier but of the whole town. The new construction was named Maritime Buildings and for over seventy years it provided office accommodation for Macbrayne's and for many other businesses, until it succumbed to time and weather and had to be demolished.

In December 1934 the full Commission decided to make application to the Treasury and the Development Commissioners for a loan of £23,500 to carry out the work. Alasdair Macleod's submission drew attention to the favourite argument of the Commissioners that all the improvements to the piers since 1865 had been paid for by loans and that the repayments of these had been funded from harbour revenue. He further pointed out that 'during the years 1914-1919 the Government took control of Stornoway Harbour as a base for the North Atlantic Patrol and used all the wharves and piers for Admiralty purposes. During that period no repairs or improvements were carried out to the structures and no harbour dues were paid, nor did the Admiralty grant any compensation on termination of hostilities.'

The offer of grant and loan was immediately accepted and Henderson and Nicol were directed to prepare the contract documents for the work.

The Motor Takes Over

The decade of the1930s saw an enormous change in transport and communications. In the fishing fleet the sail had almost completely given place to steam by the early 1930s and steam itself would soon be replaced by the diesel engine. An experiment in 1934 when the Girl Helen of Findochty and the Poppy of Buckie were launched with diesel engines, proved highly successful and was hailed as an economic proposition for the faltering herring industry. The motor lorry and the car had replaced the horse and cart on the Stornoway streets and a variety of buses had begun to wend their way daily to all corners of the island. This increase in the use of oil and petrol was reflected in the business of the Harbour Commission

Coal hulks were a familiar sight in Stornoway harbour from the 1880s into the 1950s. However, in the 1930s a number of oil hulks began to appear in the bay, the first one mentioned in the Minute being the *Constance* owned by Murdo Maclean & Sons. In April 1934 Kenneth Maclennan & Co. asked for permission to moor their oil hulk *Sweet Promise* close to the *Constance* if possible. The Harbourmaster stated that he would prefer all hulks to be confined to the area southward of the River Creed to allow the anchorage opposite the river to be kept clear for vessels seeking temporary shelter. He recommended that both the oil hulks anchor between the coal hulks and Glumaig Bay.

One of the best known entrepreneurs in the Stornoway motor trade was John Mitchell, a native of Shawbost, who in the 1930s was busy establishing his successful bus company and garage in the town. In the Spring of 1934 he applied for the tenancy of the 'small piece of ground to the right hand side of the Bayhead Street entrance to Cromwell Street Quay' on which he wished to erect a hut 24 ft by 12 ft for a Transport and Parcel Office. The Commissioners felt this was not a good place for the size of building he had in mind and instead offered him Station No 1 on the other side of the entrance, where the Resident Engineer's office was presently sited. They gave permission for a temporary building to be erected.

Mr Mitchell wasted no time in acquiring the Engineer's Office and submitted his plans to the next meeting. He intended to break a door and windows in the side fronting the roadway and to attach a small verandah, all of which were approved by the Commissioners. However, not everyone was in favour of the plan and Miss Allanina Macleod, who lived near the bottom of Kenneth Street, wrote to the Commission voicing her objections. Her point was that there was already considerable congestion at this junction, and the nature of the proposed business was bound to add to it. The value of the residential properties in the area would be affected 'due to the increased din, dust and clatter' and the increased traffic at an already hazardous corner would be a really serious menace to the pedestrians and householders in the neighbourhood.

Miss Macleod had good reason to be worried since her house had twice been badly damaged by runaway traffic. 'On one occasion a horse and lorry dashed into the ground floor window and in the summer of 1932 a heavy passenger bus dashed into the doorway, tearing away the lintel and causing considerable damage to the lobby and cement side walk.' On both occasions, she said, the occupants had only narrowly escaped serious injury.

John Mitchell's Parcel Office remained in place and by the beginning of 1935 he was asking for some improvements to another property he rented on harbour ground, this one on Esplanade Quay. The Special Committee recommended approval of his plan to roof in with corrugated iron the yard rented at the corner of Esplanade and North Beach Quay and asked for a lease of the premises instead of an annual rent. Commissioner Maciver thought this would increase the value of the Commission's property, although they must make sure that the Commercial Hotel did not lose any light by it. Mrs Fraser agreed, saying that 'work of this sort ought to be encouraged.' He was given a five year lease of the area at a rental of £26 per annum.

However, in January 1937, an application which dwarfed John Mitchell's 2000 gallon tank appeared on the agenda of both public bodies. Messrs Scottish Oils and Shell Mex Ltd already had a yard at the corner of South Beach and James Street and they now wished to erect four large tanks and two small ones with a total capacity of 900 tons in this yard for the storage of petrol, kerosene and diesel oil. This was a matter for the Town Councillors but the company also wished to berth a tanker on the west side of No 3 Wharf to lay a valve box and run three 6 inch pipelines under the surface of the quay and the harbour property to the proposed tanks in James Street, which would allow them to establish a bulk Motor Spirit and Oil Storage Depot.

The Commissioners referred this request to a Special Meeting but, in the meantime, the Secretary wrote to the Company suggesting that they consider the merits of No 2 Wharf, rather than the elderly timber structure of No 3 which might not be able to cope with the extra traffic this new operation would entail. Shell Mex at first considered sending their Marine Expert to look at No 3 but later decided against this, saying that extending the pipeline to No 2 Wharf would be too costly. Their shipments to Stornoway would be comparatively infrequent as they proposed 'providing themselves with generous storage which would mean that No 3 Wharf would adequately meet their requirements for several years to come.'

Bailie Tolmie said that if the company wanted to run the pipeline to No 3 Wharf, the Commissioners had no option but to let them have it. Provost Smith took a different view:

> 'As Harbour Commissioners we are concerned only with the application to lay a pipe line from the pier to James Street, provided they are allowed by the appropriate licensing authority, to put 900 tons of petrol or oil in this very dangerous centre of the town, in proximity to the gas works, in proximity to the mills, and in proximity to a residential quarter. Such a position does not exist anywhere in Great Britain, without great precautions being taken, and even where these precautions are taken, great danger exists. If anything happened there, it would wipe out, not only the neighbourhood, but practically the whole town.'

The Chairman, seconded by William Mackenzie, moved that they grant the application and Alasdair Mackenzie said he would support it provided there was a clause incorporated in it that their permission was granted subject to certain safeguards to be set out in a Minute of Agreement. The Chairman's casting vote went to the motion and the matter was decided. The discharge of bulk oil cargoes remains in this area of Stornoway harbour to the present day.

In May 1937 William Grant wrote a note in the inside cover of the Arrivals Book which stated: 'On Tuesday 17th May 1938, MV *Shelbrit 3* was the first vessel to discharge oil in bulk into the recently constructed tanks in James Street.' The London-registered vessel arrived from Ardrossan and departed for the Manchester Canal on the same day.

Seaplane service

The increase in the use of the aeroplane for travel and transport also had its effect in the islands in the 1930s. Stornoway Harbour had a number of visits from seaplanes from the mid 1920s onwards, and by 1931 Thomson's Tours were considering a seaplane tour of Scotland which would take in Stornoway as well as Wick, Oban and Helensburgh. A daily air service between Inverness and Orkney began in 1933 and the air companies began to investigate the possibility of a connection between Inverness and Stornoway. At first it was considered that ordinary aircraft would not be able to land at Stornoway and that any service to the Hebrides should be by seaplane but in May 1934 Stornoway Trust applied to the Land Court to resume an area of common pasture for an airport.

At one of their Special Committee meetings that month, the Harbour Commissioners decided it was time to take some action on this potential trade themselves and agreed to communicate with any interested 'aviation companies' who were interested in running a seaplane service that the Commission would be pleased to do everything possible to facilitate any such service. Alasdair Macleod accordingly wrote to Highland Airways Ltd and Midland and British Air Ferries Ltd offering them the most favourable terms possible in terms of harbour dues.

The correspondence is merely noted in the Minutes but the *Gazette* reported the response from Highland Airways as stating that they 'were unaware that they would have to pay anything for the privilege of landing on the water inside the harbour. They would have to arrange a boat to meet the plane and if they were going to be charged over and above that they would have to make arrangements to utilise a loch inland.' The Commissioners replied that under the terms of the Stornoway Harbour Order 1926, they may be obliged to levy a charge on aeroplanes but 'they would be as considerate as they possibly could.'

The decision to construct the airport at Melbost effectively ended any possibility of a regular seaplane service into Stornoway and the matter dropped from the Commission agenda. However, in June 1934 *The Scotsman* reported that it was highly likely the port would become a seaplane base for the RAF's North Atlantic route. The flying boat *Perth* had visited the harbour three weeks earlier to carry out a survey and a squadron of seaplanes from Oban also made an overnight call on the way to Kirkwall. The paper added that 'permanent moorings for seaplanes will shortly be laid in the harbour' and a few years later this type of craft did become a familiar feature of the harbour scene during World War II.

The idea of a commercial venture was still being investigated in May 1937, when West of Scotland Air Services Ltd proposed to run a service from Greenock to Stornoway via Oban, Tobermory and Portree. The Commissioners considered a letter from a Captain Glyn Roberts informing them that, if a seaplane service was to be established in Stornoway, it would be necessary to obtain a licence from the Air Ministry for the Outer Harbour to be used as a commercial landing ground. Captain Roberts said that he intended to make the necessary application and for this he required the permission of the controlling authority.

Bailie Tolmie said that they should give the company every assistance possible because it would mean a great deal to the town and island. The Harbourmaster reported that he had gone over the harbour with Captain Glyn Roberts and they had settled on Glumaig Bay as the most suitable place to moor a raft. The Commissioners gave their permission and also directed that Messrs West of Scotland Air Service be informed that 'it is the Commission's desire that, in any scheduled air services by any Company embracing Stornoway these should exclude Sundays for the landing and departure of planes under any circumstances except for cases of necessity and mercy.' On a more materialistic note, the Secretary was instructed to remind the Company that it might be necessary to exact dues for seaplanes using the port, but that these would be nominal. To this end, another of the Secretary's tasks was to find out what other harbours did in the same circumstances.

Fishery matters

Throughout all the construction work and political manoeuvring, the fishing trade continued to go about its business, bringing in the revenue on which the Harbour was so dependent. The landings were nowhere near those of the boom years of the 1920s and the industry continued to see-saw between good catches and bad markets. The Commission saw their income drop from time to time according to the state of the trade.

The advent of the trawler in the Minches caused great disturbance, sweeping clean the fishing grounds and leaving little for the inshore fishermen. Vessels with blacked-out fishing numbers, their crews' faces obscured by mufflers or handkerchiefs, sometimes came so close to the shore that the crew's voices could be heard. *The Scotsman* reported in July 1933 that a resolution to be submitted to Government had been prepared a year earlier by Stornoway Town Council, asking for an extension of the prohibited area, an increase in penalties for breach of the law, and the provision of more and faster Fishery Cruisers. This resolution had been adopted by the Town Councils of Tain, Golspie, Fortrose, Nairn, Oban, Wick, Rothesay, and Peterhead, as well as the District Committees of Caithness (South), Gairloch, Lochbroom, Ross-shire (South-West), Harris and Barra, among others.

A meeting of the Lewis Fishermen's Association in February 1935 refused a request from the Mallaig Herring Trade Association to adopt the ring-net method of fishing for the Stornoway area. The fishermen believed this method to be 'detrimental to the industry' A *Scotsman* article described the Stornoway fleet that month being in a worse state of repair than the east coast one.

> 'Most of the local motor boats were bought second-hand. They had completed their useful life for fishermen working from all ports in all weathers, but were still serviceable for fishermen operating only from Stornoway in the sheltered waters of the Minch and that only for part of the year. But deterioration has been rapid. . . If a rigid standard of age and seaworthiness was adopted for Scotland, the Stornoway fleet would be reduced quite disproportionately – indeed only a handful of steam drifters would get through the sieve.'

The Harbour revenue took a drastic drop in 1934, with the dues on both fishing boats and herring landings fell by 50 per cent. The dues on freshed herrings in 1933 were £260; in 1934 they totalled £4. Dues on kippers, on the other hand, doubled in the same year. In 1935 the Herring Fishery Board was established to regulate the industry and received a cautious, even nervous, welcome. The May fishing that year had a bad start owing to the markets being swamped with Norwegian herring under the terms of the agreement between the two countries and the uncertainty of the German market. It was frustrating because there was plenty, good quality fish to be caught that year. In early June the Stornoway market was glutted by 5000 crans of herring landings, three times more than the curers could cope with at one time, and the port had to be closed. A few days later the landings were small with only three boats catching anything at all, and two days after that, 115 boats landed over 2000 crans. The Herring Industry Board set up Area Committees in Peterhead, Fraserburgh and Stornoway.

The winter herring fishing had a bonanza in December 1936 when a large shoal appeared in the harbour. The *Easter Rose* was the first boat to come on it, on Monday night 7th December, when she was leaving for sea. Seeing the signs of herring she shot her nets in Glumaig and netted 100 crans. The *Lews* and the *Ellen and Irene* also took a good catch. The *Gazette* reported that the news went round like wildfire the following day. The paper quoted an eyewitness as saying 'a net could be seen coming out of every second door in Newton' and said that almost anything that could float went to sea that night.

> '. . .on Wednesday the market had to deal, not only with the landings of the drifter fleet but with innumerable small catches landed by skiffs working with one or two nets. At night the herring could be seen playing close in at the pier, in the light of the electric lamps. Some youngsters with a torch caught quite a good fry on the shore at South Beach, by the simple expedient of wading out and kicking the herring ashore; at Newton quite a number of herring were left high and dry by the receding tide.'

But the fishing industry faced another crisis in May 1937, this time from outside its own shores, when the German importers announced they would not accept May herring. They said that because of the large quantities of food being stored in Germany there was no cold storage available for the fish. The Local Area Committee decided to close the port until the situation was clearer. The fishing boats from other harbours joined in the appeal to the Herring Industry Board to do something. Some of the East Coast and English boats were already on the way to Stornoway for the summer season fishing and the BBC broadcast a warning in all their news bulletins on the Saturday night that boats on passage to Stornoway should not shoot their nets because the port was closed. A German steamer arrived to pick up a cargo of herring but nothing was put aboard her at first, since the curers maintained there was no point in wasting the barrels if their contents were not to be sold or cold stored in Hamburg. By the Tuesday evening, however, they took a gamble and shipped 5000 barrels to Hamburg, in the hope that the quality of the herring would persuade the Germans to sell or store the cargo.

The kippering businesses were brought to a standstill as well, with no herring being landed. The *Gazette* reported that 120 boats lay idle in the harbour and the gutting crews had nothing to do. The fishing industry was angry with the Herring Industry Board for letting the situation develop and for not warning the industry. The timing could not have been worse for the port of Stornoway. The season had looked promising, catches had been good, there was a large fleet using the port, most of the curing firms had been back on the quays and '. . .it was generally hoped that the ten years spell of bad fishings had at last been broken' said the paper. The report added that there was much cause for uneasiness; the same quality of fish was selling in the market for 15/- per cran as had sold for 40/- per cran in the same period in 1936.

The Mail Service Again

In December 1937 Robert Steven, seconded by Roderick Smith, moved that the Secretary should make inquiries into the imminent expiry of the contract held by David MacBrayne Ltd for the conveyance of mail, with a view to making representations on the improvement of the service before a new contract was concluded. Mr Steven also suggested that they should campaign for a reduction in fares, while Mr Smith added that they should ask for a faster boat. 'This is the slowest crossing in the whole of Great Britain, where a mail service is concerned, with the exception of other services offered by the same company,' he said. The *Lochness* had been heralded as being able to cross in four and half hours, but she had never, to his knowledge, accomplished the crossing in less than five and a quarter hours and her usual time was five and a half. She was no real improvement on the *Sheila* who had frequently managed the same time and always less than six hours.

The General Post Office in London informed Alasdair Macleod that the contract would expire at the end of 1938. Bailie Tolmie opened the discussion at their January meeting by saying:

> 'We as Harbour Commissioners have been very progressive for a number of years now. We have improved our piers at a cost of £120,000 and made everything possible for the progress of the community. The goods service to the island as far as Coast boats from Glasgow are concerned is quite all right but I must say that our passenger service is very far behind.'

The criticisms became more detailed as the other Commissioners took up the theme. Bailie Mackenzie, along with Captain Macdonald and Councillor Gunn, had timed the stop at Applecross and reckoned it took half an hour; the train connections at Kyle delayed the vessel; and Bailie Mackenzie was also of the opinion that the ship could manage an extra couple of knots, if the company would let her. MacBrayne's may protest as to the cost, but under the contract the company was protected from loss and surely the Treasury would not object when they realised the benefit to the community. He recalled being on the old *Gael* in 1913 or 1914 when she did the crossing in four and a half hours. 'I think the skipper got into trouble because he had to fire the second boiler and he never did it again,' added the Bailie.

Councillor Gunn quoted the famous notice on some of the company's ships: 'THIS DECK HAS ACCOMMODATION FOR SO MANY PASSENGERS, WHEN NOT OCCUPIED BY CATTLE, SHEEP OR OTHER ENCUMBRANCES.' If the community was trying to attract tourists to Lewis, there must be a good ship and an attractive fare, he said.

Mr Tolmie was justly proud of the work accomplished by the Stornoway Pier and Harbour Commissioners in a space of less than ten years and in one of the most difficult financial climates of the century. They had spent over £120,000 on upgrading the Harbour, had taken on a total debt of £70,000 and against stiff odds, had almost completely rebuilt their quays, giving some of the best accommodation of any fishing port in Scotland, installed electricity on all of their piers and built a brand new flag-ship building which was bringing in revenue. Their capital expenditure had been high but their maintenance costs had dropped dramatically and they had plans for even more improvements.

This was all done against the background of great changes in the wider world – the Wall Street Crash, the Jarrow March, the death of the old King and the abdication of the new, the coronation of the Duke of York, the rise of Hitler in Germany, and now, in the last two years of the decade, the threat of another world war. In the middle of 1939, all the plans had to change.

Norman Mackenzie

SY725 *Dochas* on South Beach for minor repairs, probably 1949. Her Gaelic name is translated as Hope.

Stornoway Historical Society

Laying the foundation stone of Stornoway Town Hall 1903
In attendance some leading personalities of the Town Council and the Harbour Commission (l–r) John Norrie Anderson, Major Matheson, John M Morrison, Murdo Maclean, Foreman Mason, Archie Munro, Peter Macleod, Danny Maciver, Aeneas Mackenzie and Kenneth Maclennan

Mrs Mairi Maclennan

(left) Peter Macpherson, Deputy Harbourmaster for over 20 years. Possessed of a formidable voice which it was said could be heard in every creek and gully on the other side of the bay, he controlled gatherings of up to 1000 boats with ease

Malcolm Macdonald

Three Collectors of Dues: William John Macdermid 1941 (above); Angus Macdonald 1941-59 (below left); Angus Macleod, 1912-1922

M Macfarlane

A future Collector and a future Harbour Commissioner – Donald Murray and Malcolm Macfarlane in the 1930s

Stornoway Historical Society

Characters of the Commission . . .

Macleod family

Alasdair Macleod – Clerk to the Commissioners 1931-1968

Mrs S. Mackenzie

Captain George Watt, Harbourmaster 1919-24. Admiralty pilot at Stornoway during WW1 – *'an active and splendid-living man, in every sense of the word'*

Retiral of Alexander Mackay, Harbourmaster 1960-78 – (l-r) Donald Murray, Collector of Dues; Catherine Mackay, Collector's Clerk; John Macleod, Chairman; Captain Mackay; Mrs Christina Mackay; Murdoch Macleod, General Manager; Patricia Smith, Secretary

Norman Mackenzie

Fish laid out on No 2 Pier awaiting auction. The sales were always an attraction for the general public – passers-by, tourists, and small boys included

Sandy Matheson

The Lights Go Down Again World War II 5

BRITAIN'S DECLARATION of war on 3rd September 1939 put a halt to the Commissioners' improvement plans for the immediate future and brought the harbour back into military use after just twenty years. The warning signs had been appearing on the international stage throughout the 1930s, from Hitler's appointment as Chancellor of Germany in 1933 to his annexation of Austria in 1938.

At Stornoway Harbour the first indication of preparation by the authorities came as early as July 1937 when the Secretary received a letter from the Ministry of Transport enclosing a copy of the Air Raid Precautions Handbook No 7 – 'Dealing with anti-gas precautions for Merchant Shipping'.

A year later a letter dated 2nd June 1938 and headed 'Secret' came from the Mercantile Marine Department of the Board of Trade to say that the Board and the Admiralty had been discussing how many of the navigation lights on the coast should be extinguished in the event of possible hostilities. The lights on Arnish Point and Reef were to remain lit but it was for the Pier and Harbour Commissioners to decide whether or not to extinguish the navigation lights on the wharves. If the Commission did want to take this course then, in the event of a possible emergency, the Department would send them a telegram with the instruction 'Extinguish navigation lights'. If the life-saving services required the lights to assist them at a night rescue, then this would still be permitted. Alasdair Macleod's reply on 21st July 1938 said that he had consulted the Harbourmaster who considered it best to leave the lights on until otherwise instructed by the authorities.

The country came to the brink of war in September 1938 during the Austrian crisis and No 1 Pier was the scene of more mass leave-takings as the Lewis Reservists answered the call to mobilisation. Chamberlain's Munich Agreement postponed the evil hour and the men came home again, but Stornoway Town Council asked if the Commissioners would make the Battery Buildings available as an Air Raid Precautions Depot if required. On the motion of Norman Maciver, seconded by David Tolmie, the Commissioners agreed, setting a rent of £80 per annum. Captain Thomas Macdonald moved a rental of £30 but received no seconder. The Commissioners received no rent either, as the Council refused the tenancy on the terms requested.

At the same meeting in October 1938 the Commissioners noted a letter from the Admiralty informing them:

> 'it may be necessary, in view of the emergency which existed at the time of writing, to take over for naval purposes four rooms in Maritime Buildings. In that event. . . a representative of the Admiralty bearing a written authorisation will present himself at the premises for the purpose of taking possession.'

The matter of compensation would be the subject of a later communication, the Admiralty added.

The summer fishing season of 1939 seemed a precursor of the dark days to come. The herring shoals vanished from around the coast and all the fishing ports suffered from the collapse of the industry. Almost all the local boats lost heavily and the fishergirls returning from other ports came home with very little money. Preparations for war moved up a gear with 20,000 gas masks arriving at the port and Air Raid Wardens appointed for each of the eight divisions in the town. The departure of the *Lochness* that week was very busy with the visitors from the island sporting estates and other holiday-makers cutting short their annual breaks and returning south early to their various homes and responsibilities.

The British fleet was mobilised on 31st August and James Shaw Grant recorded the scene at No 1 Pier, as once more, the people gathered to send off their young men to war.

> 'The throng on the pier was tightly packed right up to the edge. From the main barrier at MacBrayne's office the crowd stretched round the end of the Fish Mart and was lost in the darkness beyond. Only a narrow lane was left through which the reservists dribbled in a slow but regular stream. At the other end of the pier, towards King Edward's lamp, there was another crowd, numbering several hundred. . .'

He watched from the roof of Maritime Buildings as the deck of the *Lochness* filled up.

> 'It was an orderly and strangely silent crowd. There was not even a hum of conversation. . . There was a steady stream of reservists

> through the crowd and up the gangway but their leave-taking with relatives was also silent. A quiet handshake, at most a single word of farewell, then up with the kitbag and away. . . not a whisper rising from the crowd.'

It was the Sunday evening of the August Communion weekend and many of those present had come straight from church to the harbour. A precentor somewhere in the throng began to give out the lines of Psalm 46 in Gaelic. As the *Lochness* moved slowly away from the quay, the reporter saw. . .

> 'a flutter of handkerchiefs from the deck. The crowd ashore poured through the barriers and rushed to edge of the quay. Only then did they find their voice. A full-throated cheer broke out from the crowded decks and waving frantically the crowd ashore responded.'

All Change

Those remaining at home had to adapt to the new circumstances. James Shaw Grant noted:

> 'Many of the Harris Tweed firms have had large orders cancelled. . . The crops are of more than ordinary importance this year and most of the young men in the rural districts have already been called away. The work therefore will fall upon the women. . .'

Some foodstuffs were rationed by the end of September and freight charges on coal and gas almost doubled within weeks. The Sailor's Home received some of the early casualties of the war in October when the crew of a torpedoed Greek steamer were landed at the port.

Matters on the Harbour Commission meeting table changed as well, and their plans were put on hold. Just a week before the declaration of war the Works Committee had been recommending putting the Fish Market plans out to tender in four mainland papers as well as the *Stornoway Gazette*. On 12th September 1939 the Commissioners were forced to suspend the scheme, 'on account of the difficulties in the matter of labour and materials under war conditions.'

The Chief Surveyor of Lands, Admiralty, London, wrote to inform the Commission that 'it may be necessary for the Admiralty to take over for naval purposes (1) Maritime Buildings, King Edward's Wharf and (2) Berthing and Anchorage for Admiralty vessels, Stornoway Harbour.' Alasdair Macleod reported to the 12th September meeting that 'Captain W D R Crowther, RN, Senior Naval Officer at Stornoway, had presented himself at 10.45am on Friday 1st September 1939 for the purposes of taking possession of the premises which he did on that date.' The Commissioners instructed the Secretary to ask the Admiralty what rent would be payable on the building and agreed to keep the matter of compensation in view, in case of deterioration or interference with the structure as a result of the occupation. On 22nd September, Captain Finlayson, who was a Lieutenant in the RNR, announced his secondment to the Navy as Second in Command at the port and the Deputy Harbourmaster, Malcolm Maclean became Acting Harbourmaster.

In November Norman Maciver suggested they should make a site available for a portable building to be used as a canteen for supplying meals and refreshments to servicemen arriving and departing on the steamer. The Ladies Committee, who intended to run the canteen, asked if, in the meantime, they could use the Harbourmaster's office. Putting their case, Norman Maciver said that although there were not many servicemen coming home yet, it was felt that those who were travelling should have a place where they could get a cup of tea or coffee before getting their buses. The Harbourmaster's office was far too small and very inconvenient, since the water would have to be taken from the Fish Mart and the ladies would be dependent on a coal fire for heating, but they were very anxious to start and this would be only a temporary arrangement. The Commissioners agreed to the request. The ladies took on the task of catering for the men arriving on the mail boat, while the YMCA remained open until after the vessel sailed, catering for those who had to spend a long day in town before departure.

Dark Days

The closing weeks of 1939 were shadowed by the news of twelve Lewismen missing after the loss of the *Rawalpindi*. At the beginning of 1940 reports began to appear in the press of the remarkable feat of 18 year old Calbost boy, Malcolm Morrison, who had sailed a ship's lifeboat over 300 miles to safety after the steamer *Arlington Court* was torpedoed in the Atlantic. A heading in the *Stornoway Gazette* of 26th January 1940 read 'A Dark Week for the Island' referring to the heavy list of Lewis casualties after just four months of war.

The Secretary reported on 9th January 1940 that a Government circular had been issued, regarding the restoration of some form of street lighting but details of the type of approved fittings had not yet been received. Ten days later, the *Gazette* was reporting the third fatal accident since the blackout began, when a body was found at Cuddy Point. 'There is no immediate prospect of any lighting in the town or any of the piers and the public must continue to suffer the inconvenience and the dangers meantime,' the paper reflected.

Rumours had begun to circulate towards the end of 1939 that the mail steamer might have to begin calling at Portree and that the service might be curtailed even further. The Commissioners had decided to join forces with the other public bodies with a view to making a protest.

At the January meeting, the subject of dues on the mail steamer came up when Norman Maciver referred to the decrease in revenue in the last financial year, caused by the short summer herring fishing season. He warned that there would be a further reduction in their income as a result of the war and that the Commission should keep all expenditure to a minimum. He suggested that, in the circumstances, the special concession granted to the mail steamer should be reconsidered, and the Finance Committee was given the task of looking into this. They recommended that, as from a date to be determined by the full Commission, the concession should be discontinued on the grounds that the settlement with MacBrayne's back in 1928 had been intended to continue until the 'necessities of the harbour forced the Commissioners to increase the rate.' They pointed out that no increase had been made when the larger *Lochness* had replaced the *Sheila* on the run. The Commissioners accepted this recommendation at their February meeting and instructed the concession should cease as from 1st March 1940.

This subject came to dominate their discussions over the weeks. MacBrayne's had reported the Commission's intentions to the Ministry of Transport, claiming that the charge for the *Lochness* was going to leap from £750 to £4,635 per year, an increase of roughly 550%. The Ministry said that they viewed this with concern and reminded the Commissioners that the Defence (General) Regulations 1939 gave the Minister certain responsibilities and powers of control over harbour charges. The Minister would therefore be glad if the Commissioners would give him their reasons for their action and in the meantime 'the Commissioners may consider it desirable in the light of this letter to defer making any change in their present charges until the Minister has concluded his examinations'.

A draft reply was submitted to the March meeting and approved. In it the Commissioners claimed that the *Lochness* dues were being over-stated and the proper amount would be £2331. They stated that the sum had been calculated on a much smaller vessel than the *Lochness* and had been applied to her and her relief vessels without prejudice for the last twelve years. They also explained:

> 'as a result of the loss of markets for cured herring, formerly available in Germany and Poland, the herring industry, on which the Harbour Revenue is heavily dependent, will not be prosecuted to any extent comparable with previous years and as a consequence the Commissioners anticipate a decrease in revenue from dues on

> herrings landed and exported, on fishing boats and general shipping traffic, as also from rentals of heritable properties and curing stations. . .'

They emphasised that the move to increase dues was prompted by their desire to fulfil their financial commitments. The meeting did, however, agree to hold off charging the increase until the Minister of Transport had dealt with the matter

The Minister refused to sanction the increase, noting that the whole loss of revenue from the herring trade, amounting to £1,500, 'would thus be borne by the steamer used by Messrs MacBrayne in carrying on part of the essential transport services in the Western Highlands and the Hebrides under their contract with His Majesty's Government.' He was, however, prepared to authorise an increase of a more general character should it prove justified and therefore a questionnaire was enclosed for completion if the Commissioners wished to apply under those terms.

The Commissioners backed off, Thomas Macdonald suggesting that they should not press the matter any further but keep it before them should circumstances change. Alasdair Macleod observed that it was unfortunate that MacBraynes had been able to invoke the Defence Regulations, adding that in normal times they cold not have opposed the increase. The Secretary was instructed to reply to the Ministry and the Collector was authorised to resume rendering the *Lochness* at the old rate.

Another dark year gone

The first full year of the war had seen huge losses for Britain. Norway, Holland, Belgium and Luxemburg had fallen to the Germans in the first few months. A National Government was formed in May, led by Winston Churchill, and the following month many Lewis servicemen were present at the evacuation of Dunkirk, among them Pte Ossian Macaskill RAOC, one time Harbour Commissioner. The *Stornoway Gazette* reported that Pte. Macaskill did 'yeoman service with a small boat ferrying members of his unit out to a waiting destroyer'.

There were fears of anti-Italian demonstrations against the small community in the town in the summer when Italy declared war on Britain, but these were unfounded. In July the list of the Missing in the *Stornoway Gazette* filled almost a full column. There were instructions on what to do in an Air Raid, and Days of National Prayer were observed in May and in September. In a single week that autumn Lewis and Harris raised a total of £6,416 16s 2d to pay for a Spitfire aeroplane.

Over the months of 1940, the Commission business had been shaped by the war, operationally and financially.

Small, almost matter-of-fact, items in the Minute Books show the signs of the time. The let of the basement of Battery Buildings, given up by the ARP, was taken on by the Town Council for storing and baling waste paper, a campaign that had grown in momentum in the community. On 29th August the Royal Air Force took over the old Magazine at Battery Point 'for the storage of bombs'.

In September the Acting Harbourmaster was instructed to have all the lamp standards, bollards and railings painted white as an aid to movement during the blackout, while in October, after correspondence from the County Clerk, the Commissioners agreed to make the Harbour Mortuary and the adjoining apartment available as an emergency mortuary should it be required. The old Employment Exchange was rented out to HM Office of Works from November at £20 per quarter for a recently set-up Medical Board.

Although the Commissioners had been afraid of running into financial difficulties because of the war, the first entry in the Minute Book of 1941 is a meeting of the Finance Committee to deal with 'Surplus Revenue'. They agreed unanimously to set up a Contingency Fund to which any surplus revenue could be put, up to prescribed maximum of £10,000.

But the Commissioners still kept a strict eye on spending, instructing the Harbourmaster to make periodic checks on the Admiralty-occupied buildings to make sure they were being kept in a good state of repair, 'in view of the reduction for repairs in the compensation rent offered.' In April the District Valuer forwarded his decision on the Admiralty compensation for the occupied areas. The offer was £702 per annum for King Edward Wharf and £211 for North Beach, more of which had been requisitioned for canteens that March. The Commissioners were not happy with the offer and asked the District Valuer to increase it, 'especially having regard to the extensive use made by naval craft of parts of the Harbour not requisitioned'. In May they received his new recommendation – £1039 for King Edward's Wharf and £234 for North Beach Quay, which the Commissioners accepted.

The occupied areas continued to extend throughout 1940; at the Annual General meeting at which they had accepted the new compensation figure, the Commissioners noted that the Navy had requisitioned the office at No 1 Wharf tenanted by Messrs Wolkoff & Son as well as those at Nos 1 and 3 South Beach and the store on Esplanade Quay occupied by Messrs Kenneth Mackenzie Ltd. At a Special Meeting a week later they agreed to rent ground on South Beach Quay to the Air Ministry Works Directorate for the siting of Nissen huts for RAF Embarkation Offices and Staff quarters, as well as temporary accommodation for 'nightly incoming personnel.' The Commissioners charged 10 shillings per week for each of the curing stations given up.

In August, when it became apparent that the RAF intended to establish permanent works in this area, the Commission decided it would be advisable to negotiate compensation directly with the Air Ministry, especially since at least part of the ground could not be returned to the tenant who had used it for agricultural purposes. The October meeting heard that their claim for £135 16/- compensation for the ground, was considered 'reasonable'. The sum was less £80 which was being paid separately for land and property at Battery Buildings.

The weekly rent of 10/- for each of the five stations occupied by the RAF on South Beach Quay was considered excessive by the Air Ministry and in October they offered an annual rent of £14 for each, the upset rental previously received on the sites. The Commissioners agreed to press for this amount to be increased and to point out that the £14 would only have been payable for a four month fishing season. This point was lost on the Air Ministry who replied that 'had the ground been left in the hands of the Commissioners, the revenue receivable would not have exceeded the sum of £14 per annum per plot offered which is considered fair and reasonable.'

A result of the Air Force presence at the Battery was the installation of a pipe line at No 3 for bulk petrol. In April 1941 the Air Ministry made inquiries about the possibility of using a wharf for the unloading of bulk petrol and, on the motion of Norman Maciver seconded by ex-Provost Smith, the Commissioners granted consent to have pipe lines installed on No 3 Wharf on the same conditions as those granted to Shell Mex. No reply came from the Ministry for some months and in August it was noted that work had started on laying the pipelines leading to the proposed depot at Battery Point. The Engineer in charge was approached and the Commissioners were told to expect the Authorities concerned to communicate at an early date.

Protecting the Public

As well as the occupation of the harbour, other consequences of the war impinged on Harbour Commission meetings. The Ministry of Transport issued the Fire Prevention (Business Premises) Order 1941 in January and, although its powers did not yet extend to Stornoway, the Secretary reported that 'it was expected that measures should be taken to deal with incendiary bombs and fires.'

The Commissioners agreed to co-operate in the formation of one or more fire-fighting parties at the harbour, to obtain more fire-fighting equipment, and to set up a Committee to deal with the matter. In March the Special Committee recommended the purchase of 600 linear feet of canvas fire hose, 4 hose nozzles and 2 standpipes, as well as a scoop and shovel for dealing with incendiary bombs. The Harbourmaster was instructed to find out the type and size of couplings on the 'donkey

Changes at the Harbour Office

For twenty years William Grant and William John Macdermid had worked together as Collector of Dues and Assistant Collector respectively. In his mid-sixties William Grant had been with the Commission for over fifty years, starting as a boy clerk whose first job was to stand on the point of the pier and count the fishing boats as they came into the Inner Harbour. Now, at the beginning of 1941, he felt it was time for him to retire and accordingly tendered his resignation as from 11th March. Although William had been born in Harris, the family had come to live in Stornoway when he was a child, settling in Bayhead where William grew up. He never married but lived out his days in the family home where his sister Maggie ran the little shop well-known to generations of Stornoway children. He was a quiet man with a pawky sense of humour, who loved fishing and bowling and relished being able to walk in the Castle Grounds which had been out of bounds to the public in his youth. Seemingly never to be as much in conflict with his employers as some of his predecessors had been, the Commissioners acknowledged his 'faithful and valuable service' and granted him a bonus of £100 on his retiral.

On the motion of Norman Maciver seconded by Donald J Maciver, the Commissioners agreed to advertise the post at £220 per annum plus £15 War Bonus, in the *Scotsman*, *Glasgow Herald*, *Aberdeen Press and Journal* and *Stornoway Gazette*. They received 19 applications, 14 from the mainland and 5 from Lewis, and at their next meeting decided on a short leet: John Forrest, Glasgow, A W Wilson, Loanhead, James Smellie, Motherwell, Angus Macdonald, Westview Terrace, Stornoway, and William J Macdermid, who had been the Senior Clerk in the Harbour Office for some years. The list was further reduced to the two local applicants and the post went to Mr Macdermid by four votes to three, subject to his finding surety of £500. The Chairman suggested that they promote Donald Murray from

Junior to Senior Clerk. The Secretary said that 'Murray had been getting on very well in the office and took a great interest in his work.' The promotion at an extra £2 per week was agreed to unanimously and since he was due to be called up for military service shortly, they agreed to take on another boy to learn the work before he left. They also agreed to keep his post open for him.

William John Macdermid's tenure as Harbour Collector was sadly very short. He had joined the staff of the Harbour Office as a Junior Clerk around 1920 and was promoted to Assistant Collector in 1922 when he was about twenty years old. He married Rachel Macdonald in 1927 and they had two children but lost their son Raymond in infancy in 1937. In July 1941, just four months after his appointment, he tendered his resignation on the grounds of ill health, thanking the Commissioners for their courtesy to him during the 22 years of his service with them and offering to help his successor with any information he might need.

William John appears to have been something of a Latin scholar, judging from the elegant copperplate quotations entered on the inside covers of one or two or the 1920s Arrivals Ledgers and initialed 'WJM'. One of these reads *'Tempora mutantur, et nos mutamur in illis'* which he translated below as *'The times are changed and we are changed with them.'* The entry is not dated so there is nothing to hint at what prompted it. After leaving the Harbour Commission, he took his family to the mainland where he joined up as a private in the Royal Army Pay Corps but he died in a military hospital, just a year after leaving his home town.

The Commissioners decided to offer the Collector's post to Angus Macdonald who had been runner-up a few weeks before. 'Angie Doll' was then in his late forties with a long career in administration behind him, beginning at the age of 14 with a stint in Stornoway branch of the Bank of Scotland. After World War I service, still in his mid-twenties, he was offered a post with the British Bank of South America, first training in their London branch and then spending some years in Rio de Janeiro until the bank came under American control. He spent a year on the journey back home, taking his time and seeing some more of the world before returning to Stornoway to marry, settle down and start work as Secretary/Treasurer to the Lewis Hospital Board. He accepted the Commissioners' offer and remained Collector of Dues for almost two decades, his office becoming a daily call for the regulars on their 'constitutional' round the harbour.

Between the two appointments, Donald Murray seems to have kept the office going and the Commissioners granted him a £10 honorarium at the next meeting 'in respect of the added duties and responsibility which he shouldered and carried out so well recently'. His call-up came in September and he left the office for service in the Royal Air Force on 4th October 1941. The new Clerk whom he had been tutoring, Donald Macleod, Tong, was promoted to Senior Clerk at £1 10s per week.

The Acting Harbourmaster was granted an increase in his wages in view of the increased responsibility, long working hours and higher cost of living since the outbreak of war. One of the two watermen left the harbour staff and since no one could be found to replace him, the remaining man was granted an extra ten shillings per week to compensate for the longer hours he worked. As Norman Maciver remarked at that July meeting, the labour situation had got worse and the Commissioners, commenting on the fact that unskilled men were earning big wages even on temporary jobs, agreed that to keep their staff they would have to pay the going rate.

The new Collector advertised for another clerk and appointed Miss Nan Maclean, Keith Street to that post at £1 5s per week. Councillor Gunn had tried to have a higher wage set for this post, saying that the Commission was not a poor body and could afford to pay it. He received no seconder, ex-Provost Smith replying that 'We are not a poor body but we are a public body and we should not be in competition in these matters with private businesses that have to obtain comparative labour.' Donald John Maciver reminded the Commission that the boy whose position was being filled had started at £1 per week. 'Times are changed', reflected the Chairman. ❑

water pumps' on drifters and other vessels, so that these could pump sea water through the fire hoses if the on-shore water supply failed. Two stirrups pumps were already installed in the Harbour Office and both the steamer companies, MacBrayne's and Coastlines, were forming fire parties. The Committee instructed that another fire party should be formed from among the Commission's staff and that all three parties should co-operate under the direction of the Harbourmaster to operate the equipment.

In February 1941 the Town Council asked for consent to erect three public air raid shelters on harbour property, one on South Beach, one on Esplanade Quay, and the third on North Beach. The Council said they had decided on these sites having in mind 'the number of people usually employed on business about the piers and the need for providing some measure of protection for the crews of fishing vessels and other vessels.'

The Navy also asked approval for the construction of an Air Raid Shelter on the east of the roadway approaching No 1 Wharf, which the Minutes said, 'while primarily for service personnel would also be made available for civilians caught in an air raid in the vicinity.'

At the Commission meeting in July 1941 the Chairman, Thomas Macdonald, reported that the Civil Defence Authorities had been worried for some time about the 'vulnerability of the fishing fleet while tied up in the Inner Harbour.' The Lord Lieutenant of the County had told the Chairman that, if the Commissioners wished, he would make an urgent request to skippers and crews of fishing vessels 'recommending that they keep afloat whenever possible and to spread out as far as practicable so as not to present a compact target which might tempt the enemy into action, leading to the loss of valuable lives and vessels and to irretrievable damage to a source of the nation's food supply.'

The Commissioners agreed with the terms of the proposed notice but felt they could take no action in view of the Admiralty's almost continuous use of the deep water berths and the amount of quay space under naval control.

The military presence in the port was putting pressure on harbour users in other ways than just the berthing of ships. MacBrayne's were experiencing considerable delays in the turn round times of the mail steamer owing to the lack of storage accommodation for heavy cargoes. They wrote in April 1941 to ask if they could be provided with a temporary shed but the Commissioners replied that in the present circumstances they were unable to help. MacBrayne's held off until September when Mr Livingstone, the local agent, wrote again to say that if additional storage was not provided, the Company would have to cut down on shipments.

Ex-Provost Smith, seconded by Provost Mackenzie, moved that they suggest to Mr Livingstone that he approach Mr John Mitchell to see if they could arrange a transfer of the tenancy of the yard on Esplanade Quay. If this was agreed, the Commissioners would consent to the erection of a temporary store being erected in the yard.

Towards the end of 1941, the Commissioners also granted permission to the Red Triangle Club to extend their premises 'to meet the increased demand made on the Club at the present time by service men.' The members expressed some concern about what facilities were being provided for the local youths who had supported the club all along, but Edwin Aldred said that their interests were being catered for. The local YMCA Committee had not agreed to the suggestion that the whole of the Club should be given over to the services, he said.

Protecting the Port

The year of 1941 had seen a deepening of loss and of hardship. Cured herring was in short supply in the Highlands, partly due to the doubling of prices, but mainly to the demand in the rest of the country for fresh or kippered herring. In January the Town Council had appealed to Lord Woolton, the Minister of Food, to make some arrangements to combat that situation in the next year and allow the crofters to buy a quantity of herring at reasonable prices. Lord Woolton regretted that he was unable to do that. *The Scotsman* article which reported the Town Council's appeal, also stated that 'one well-known Stornoway firm, which alone supplies about five thousand barrels each year to crofters, is at present investigating the possibility of getting suitable cured herring from Iceland to ease the crofters plight.' Even if this proved possible, the paper added, it was unlikely that the crofters could afford to pay for the herring 'as the price will rule at considerably more than twice the pre-war level.'

Freight charges had also shot up in the first two years of war. The Town Council appealed to the Government on this front as well, stating that freight on bulk commodities such as coal and gas had increased 500% since the outbreak of war. Consequently the price of coal in Stornoway had gone from 44s per ton to 70s per ton and gas prices had increased by more than two shillings per cubic foot. Rationing on most commodities was in force by the end of 1941.

The lists of the dead and missing continued to grow as the theatre of war spread wider and wider throughout the globe – North Africa, Greece, Russia, the Middle East, the Far East – and at sea, where so many Lewismen served, the losses were heavy. HMS *Hood* was sunk in May, the *Ark Royal* in November, and the *Repulse* and the *Prince of Wales* in December. The merchant fleet suffered regularly as the convoys across the Atlantic and up to the Baltic ran the gauntlet of German U-boats. On 7th December 1941 the Japanese bombing of Pearl Harbour bought the United States into the war.

At their 9th December meeting the Commissioners considered a circular from the Ministry of War Transport which referred to plans prepared by the Naval and Military Authorities in conjunction with the Harbour Authority 'for immobilising the port for a given period by dismantling or removing essential parts of port equipment etc to meet the situation which might arise if the enemy landed in this country.'

The Minister would be glad, said the circular, to receive an assurance that the Harbour Authority were perfectly satisfied as to what they are required to do to carry out the agreed plans. Since the Harbour Commissioners knew nothing of any plans that the Navy had, Alasdair Macleod was instructed to make the Senior Officer in Charge aware of the terms of the circular and ask him to disclose the action the Navy proposed to take should such a situation arise. The Secretary stated that, from the information received in another Ministry circular, it appeared that the only action necessary at Stornoway would be 'the immobilisation of water supplies, vital parts of oil pipe lines and the sinking or removal of buoys.'

The Naval Officer's reply, dated 12th December 1941 and marked 'Secret' was submitted to the January 1942 meeting. Its contents detailing the immobilization scheme were reported to the Commissioners but not recorded in the Minute Book, nor is any trace of it on file in the Harbour Record. As the *Stornoway Gazette* explained to its readers that month:

> 'Our reports from the Harbour Commission are becoming shorter and shorter, not because the Commissioners have less to do or are any less active but because most of the business connected with British ports today falls under the censor's ban and relatively few of the matters discussed can be reported in our columns.'

From 1st March 1942 the Outer Hebrides became part of Protected Area No 4, which required any non-resident entering the area to have a military pass unless they had an exemption from the authorities. Using binoculars, telescopes and cameras in a public place also required a permit and later that year a Wren was admonished at Stornoway Sheriff Court for taking photographs of the pipe band during the Warship Week parade in the town.

Out in the wider world Singapore and Java had fallen to the Japanese, the RAF began bombing raids on German cities, the German advance eastward was halted by the Red Army at Stalingrad and the Allies defeated Rommel in Egypt. Although the tide of the war appeared to be turning, peace was still a long way off.

One of the items at the Commissioners' December 1942 meeting was yet another notice of requisition, this time from the War Department for areas required for the establishment of machine gun posts on harbour property. The Chairman and the Secretary had discussed this matter with the Garrison Commander who had agreed not to place these sand bag enclosures on the quays where they would interfere with traffic.

The beginning of 1943 saw another run-in with the RAF over the berthing of their marine craft. The Acting Harbourmaster, Malcolm Maclean, reported on correspondence with the Officer in Charge of the craft and the Commanding Office of the RAF unit in which there were complaints about herring drifters 'being allowed to manouevre in the inner harbour to the endangering and hemming in of the RAF pinnaces.' The Office in Charge held the Acting Harbourmaster responsible for damage sustained by the pinnaces in December 1942.

Good enough for a girl!

In 1942 Angus Macdonald asked for an increase in salary for the junior clerk, Ann Maclean, who had been with them for six months and whom he found willing, conscientious and capable. The Commissioners agreed to raise her salary by five shillings per week, although the Chairman thought 25 shillings was 'quite sufficient for a young girl starting out.'

Her father Bailie John S Maclean, took no part in the discussion but, when the matter was disposed of, he could not contain himself any longer. 'You are not very consistent, Mr Chairman,' he said, pointing out that Captain Macdonald had himself recently moved a war bonus for the other clerk. 'He was a boy,' replied the Chairman, 'and I do not consider that I was inconsistent.' The absence of the young men in the forces meant that increasingly the offices in the harbour area were being staffed by women. Nan Maclean had been the first female clerk in the Harbour Office and on her resignation through ill-health Christine Mackay, New Street, was engaged temporarily as a junior clerk, along with Mrs Douglas, Keith Street, who would take on extra hours when Miss Mackay left. ❑

Malcolm Maclean had replied that pinnaces had been repeatedly warned against remaining in the inner harbour and asked the Officer again to have them removed to 'suitable buoys preferably in Glumaig Bay where they would be free from all traffic and ready to proceed to sea when required and without hindrance.'

On the Harbour Front

The Acting Harbourmaster, Malcolm Maclean, continued to oversee the day-to-day operations of the port, apparently brooking no nonsense from any of the users, military or otherwise. He asked for an increase in wages in February 1943. 'Since my wages were last increased,' he wrote 'there has been a steady rise in the cost of living. The extra duties and responsibilities have been very exacting and I hope I have carried out these to the satisfaction of the Commissioners. . .'

The Commissioners agreed and doubled his annual honorarium to £104. A few weeks later, however, they were noting Mr Maclean's absence from Harbour Commission meetings and informing him that he was expected to be present unless definitely detained by other duties.

The Harbour Collector, Angus Macdonald, also asked for a re-consideration of his salary, in view of the continuing increase in the cost of living and 'increased revenue brought out in the accounts for the year ending November 1942.' On a vote his salary was raised to £300 per annum while his assistant's weekly wage went up to £2 10s.

Restrictions and regulations continued to frustrate the herring trade. Early in 1943 all but one of the firms buying fish at Stornoway had been charged with breaking the Regulations of the Fish (Port Allocation Committees) Order of 1942. The problem was caused mainly when a boat had a larger catch than the crew had estimated to the allocation officer. The authorities' interpretation of the Order was that this should be reported and deducted from the buyer's next allocation. The custom at the port, on the other hand, was that the last buyer taking herring from a boat took any surplus or suffered any shortage, a system which worked well in peace time. This was where the contraventions were alleged to be and when Duncan Maciver Ltd, after a long hearing, were convicted and fined £10 on two counts, the other firms accepted the inevitable and changed their pleas to guilty.

According to *The Scotsman* of 18th March 1943, the Port Allocation Officer who was chief prosecution witness, stated that he had on several occasions asked Messrs Duncan Maciver Ltd to take balances of unwanted catches or small shots of herring landed late at night when there were no other buyers available. The paper added that the Officer, asked if it was helpful to him that Mr Maciver had taken the fish, replied, 'Certainly. It was saving fish.'

In June 1943 the Government cut the price of herring from 98s to 91s per cran, prompting a flood of protests from the Scottish fish trade. Over a thousand fishing boats around the coast, including the Stornoway fleet, tied up for a few days and a deputation from one of the north-east ports went to London to lobby the civil servants.

In July ex-Provost Louis Bain fell foul of the Ministry of Food Order which laid down that a fish curer was obliged to charge the same price for herring which had been split, gutted, cleaned and dipped in brine, as for herring merely sprinkled with salt. Both Mr Bain and the buyer, JM Davidson of Glasgow, were fined on nine charges under the Fish (Maximum Prices) No 2 Order 1942. The practice of splitting herring was one which was peculiar to Stornoway, having developed as a means of saving labour and ensuring that the herring arrived at its destination in a better condition for kippering. This practice had not been taken into account when the Ministry made its Order but this did not change the finding of the court.

A Little Forward Planning

Throughout 1943 the tide of the war was turning. The Germans were on the retreat from Russia, Sicily was taken by the Allies and Mussolini was overthrown in Italy. In the summer of 1943, the military scaled down slightly their presence at Stornoway Harbour. In August the Admiralty handed back the keys of the offices 19 and 23 next to the Fish Market, as well as those of No 3 South Beach Street. The offices at No 1 South Beach Street and No 20 next to the Fish Market were to be derequisitioned within another few days and the Commissioners immediately set about re-letting all the premises. The offices on the quay went to the fish trade – No 19 to the fish curing firm of Duncan Maciver Ltd; No 20 to the fish merchant, J A Mackenzie; and No 23 to the fishsalesmen Messrs Bloomfields Ltd – while No 1 South Beach Street was let to Miss Margaret Maciver for the local company of Rangers. At the end of the year, the Air Ministry de-requisitioned the portions of the plots numbered 57, 58 and 59 on South Beach, which had never actually been occupied, although there had been an 'excavation' for an intended building and the Commissioners claimed the £2 10s cost of restoring the ground.

It was agreed that No 3 South Beach should become the Harbour Collector's office (which it would remain until the mid 1960s) and the Free Canteen for Servicemen was moved to the Collector's Office as soon as that was vacated.

At the close of the September 1943 meeting, the Chairman suggested that perhaps they should begin to consider post war planning for the Harbour and it was agreed to set up a special meeting for this. The Town Council had already advised the Commissioners of their decision to approach the Ross and Cromarty County Council as Planning Authority to take all necessary steps in any planning survey to 'prevent in the town of Stornoway and district housing and residential areas being encroached upon by buildings and other erections, except shops and offices, projected for industrial and commercial purposes.'

In December the Lewis Association sent them a copy of their Report on the Town Planning of Stornoway. The Association had been formed in January that year by a number of local business and professional men who were concerned for the future of the island, particularly in the aftermath of the war. It was non-political and its objects were to 'forward the interests of all the people of Lewis and to initiate schemes for their social and economic welfare.' Its Economic Survey, published in December 1943 stressed the need to revive the Lewis fisheries which were now slipping out of local hands into those of mainland fishermen and boat owners, owing to the decay of Stornoway's fleet.

The Commissioners began the new year of 1944 with a fresh determination to get back to their pre-war programme of improvements. They held an informal meeting on 18th January 'for the purpose of discussing and exchanging views in regard to Post-War Planning of the Harbour Undertaking.' Present were the Chairman, Thomas Macdonald, ex-Provost Roderick Smith, Bailie John S Maclean, Edwin Aldred, Norman Maciver, Donald J Maciver and Donald J Macdonald. The Secretary, Alasdair Macleod, noted the result of their long discussion and the priority given to each scheme:

First: The reconstruction of No 3 Wharf in reinforced concrete;

Second: The extension seawards of King Edward Wharf; and the dredging of the inner harbour;

Third: To meet any future needs for additional harbour facilities, a possible development on the lines shown in plans prepared by the Lewis and Harris Development Company Limited, in the form of a Causeway from Battery Point to Goat Island and the construction of piers and quays and the execution of dredging.

These priorities were referred to the Consulting Engineer, Archibald Henderson, for his views, but the Commissioners also had other ideas for him to investigate. With an eye on the possible development of air travel, they wanted him to advise them on what facilities might be needed for seaplanes and the best way to provide these at Stornoway Harbour. Looking to the needs of the fishing fleet, they also asked him to report on the provision of a suitably sized dock or slip in either the Bayhead or Newton areas of the Harbour.

They then moved on to discuss their plans for the buildings on the quays, setting as their first priority the erection of the proposed Warehouse and Harbour Collector's Office on King Edward Wharf, which had been postponed by the war. The second part of their scheme required much more drastic action. To accommodate their brand new Fishmarket they proposed to demolish the block of buildings 'bounded by Point Street, South Beach Street, Quay Street, and Esplanade Road to the North, South, East and West respectively.'

The meeting also agreed on the demolition of the block bounded by North Beach Street, Point Street, Quay Street, and Esplanade Road, on the site of which they intended to erect one or more new buildings. These were to be set back to allow 'the extension of North Beach Street at its existing width into Esplanade Road' as well as a footpath and double carriageway on Esplanade Road, and the widening of Quay Street. As soon as was practicable the Mortuary on Cromwell Street Quay and the adjoining public conveniences for workers on the curing stations and quays were to be demolished and a new 'convenience for females' was to be built in the vicinity in conjunction with the provision of a public convenience by the Town Council.

Turning to the site on the corner of Quay Street and South Beach Street next to the Caledonian Hotel, the possibility of using this to site a Welfare Institution for Seamen and Fishermen was discussed and it was agreed to advise the Ministry of Labour and National Service who were considering the establishment of such a centre, that this site was available. The alternative of locating a new Customs House in the southern part of the site was noted. Finally the meeting recommended that a Planning Consultant should be engaged to prepare a drawing on the first and second proposals. Someone, un-named in the Minute, mentioned Robert Hurd, a well known Scottish architect, as a suitable person for the work.

Preparing for Normal Business

At the beginning of 1944 there had still been some evidences in the Minutes that a war was being waged. In January 1944 Norman Maciver told the Commissioners that the Officer in Charge of the RAF Camp on South Beach Quay had indicated to him that it would be desirable to have the gates at the accesses locked from sunset to sunrise 'to prevent unauthorised persons from obtaining admission.' The Commissioners did not object.

Bailie J S Maclean wondered if the Commissioners made up the difference between the normal wages and the Service pay of any of their staff on war service. The only one in this position was Donald Murray, now serving with the Royal Air Force in India, and enquiries revealed that, as a corporal, he should be earning around seven shillings per day excluding any allowance for service abroad. The Minutes record that 'the service wages being higher than those received while in the employment of the Commissioners, the question of supplementing Corporal Murray's service pay did not therefore arise.'

By Spring-time the emphasis was changing back to peace time matters. On 24th April 1944, Captain Finlayson wrote to say that he had applied to be released from his Royal Navy service and that he would be ready to resume his former duties as Harbourmaster about the beginning of June. Three months later, on 1st August 1944, Captain Finlayson was back at the helm, where he would remain for another 16 years. His salary went up to £400 and the weekly wage of Malcolm Maclean, now back to Deputy Harbourmaster, was set at £5 10s.

The first report by Captain Finlayson was written the day after he returned to post. He wanted the buildings not earmarked for reconstruction to be painted which had not be done since the outbreak of war. He was instructed to try to do this by direct labour since it was difficult to secure tenders for painter work at the time.

The 'end of war' was now more than an expression of hope. 1944 had begun with the Allied landings at Anzio and the gains had continued steadily over the months, with the D-day invasion of Europe on 6th

June, the liberation of Paris in August and the landings in Belgium the following month. The Port and Transit Control Circular No 1095 from the Ministry of War Transport which arrived on Alasdair Macleod's desk in January 1945, announced that the anti-invasion preparations may be dispensed with completely and that the Invasion Instructions given could now be regarded as cancelled.

Although victory had not yet arrived and the loss and grief continued for many families, the question of what life would bring after the war began to occupy the thoughts of many servicemen and women. The *Stornoway Gazette* ran a series of letters from some of them describing their hopes – and fears – for the future. One of the hundreds of Lewis girls working in mainland factories wrote at the beginning of 1945 to say:

> 'All of us, whether in the services or not, are tired of being pushed round and when we get a moment to spare we wonder what we're going to do when the war is over. There is always the fear of unemployment or being thrown into domestic service. . .'

The 1944 report of the Herring Industry Committee had proposed a scheme of grants and loans for replacing fishing boats, an idea which was warmly welcomed in Stornoway, and in January 1945 *The Scotsman* reported that the well-known tweed manufacturer, James Macdonald, who was a founder member of the Lewis Association, had purchased two seine net boats with a view to reviving the white fish industry in the island. The two vessels, the *Daffodil* of Peterhead and the *Lilac* of Anstruther, were crewed by island fishermen and James Macdonald maintained that some of the younger fishermen were interested in going over to this method if they could access reasonable grant and loan funding.

A fortnight later the *Gazette* reported that the Board of Trade was allowing the tweed mills in Lewis to bring their spinning machinery back into operation for the first time since 1941 'in an effort to meet the needs of discharged service men trying to re-establish themselves in the Harris Tweed trade.'

The winter gales of 1944-45 had damaged the structure of No 3 Wharf so much that Captain Finlayson was convinced it was no longer safe to berth vessels there. Many lengths of timber were now missing, presumably lying on the bottom where they would be a danger to vessels berthing alongside at low water and he advised the January 1945 meeting that Scottish Oils and Shell Mex Ltd should be warned so that they could make arrangements to have their oil tankers discharged at No 1 Pier, 'possibly by arrangement with the RAF for the use of their pipes.'

Now the War is Over. . .

Within a week of the AGM in 1945, the war in Europe had ended. Through the Spring, the harbour had been coming back gradually into the Commissioners' control, as the naval authorities derequisitioned their occupied areas.

The Commissioners had several temporary huts to dispose of, which had been erected at the Battery by the Air Ministry's Contractors, and it was decided to sell all these by public roup, with the exception of the large one which was being used as a sawmill. A night watchman had to be employed for two weeks to ensure, according to the *Gazette*, that the huts would still be there on the day of the roup, since planks were being removed steadily under cover of darkness. The sale raised £203 7s 6d after deducting £19 2s 6d expenses, including the said Watchman's wage.

In the matter-of-fact reporting of Harbour business there is no hint of the sense of expectancy and relief that there must have been the Secretary's office that week. On 4th May 1945 the German forces in northern Germany, Holland and Denmark surrendered and on the 8th the war in Europe was ended. There is no mention of this in the Minutes, nor of the volley of fireworks from the vessels in the harbour as midnight struck and they welcomed in VE Day. Up on the War Memorial hill the official Town Council bonfire was unofficially lit by the exuberant youth of the town, unable to wait any longer, while another dozen or so could be seen burning around the district, on Newton beach, at Arnish, and in the villages of Sandwick and Laxdale.

During the day the town buzzed with townsfolk and celebrating airmen, the numbers swelling in the afternoon when the passengers of the country buses spilled in to add to the laughter and song echoing in the streets. Everywhere there were flags, on the Town Hall, on the shops, and on private houses, and every vessel in the harbour was dressed overall. Music was relayed from loudspeakers on the rescue boats and from the cinema. When the *Lochness* entered the harbour that night all her flags streamed in the breeze and out on deck two pipers, one a civilian, the other a sailor, played her in to the quay. A large crowd waited on the pier where almost six years earlier another gathering had watched her slip out into the darkness, taking their men away.

After the years of enforced darkness, the town was ablaze with lights. Fixed to the Town Hall was a vast flashing V Sign, erected by the Stornoway Electric Company, along with red, white and blue lights flashing the morse code victory signal. The ships in the harbour used everything they could think of to keep the sky lit until well into the night – rockets, Very lights, maroon squibs, tracer bullets, flares and searchlights, according to the *Gazette* report. Music from a loud speaker in the Council Chambers kept the people dancing in the street until one in the morning.

Many others marked Victory in Europe in church, offering thanks or reconciling their grief during services held over the two day period. James Shaw Grant, who recorded the end of the war as he had recorded the beginning, in his own descriptive, perceptive way, struggled to find words for his editorial that week. Reminding his readers that the conflict was not yet completely over and that there were still many Lewismen serving with the Royal and Merchant Navies in the Pacific, he asked the question, 'If the cost is so great, how great must be the prize?'

He answered himself with a quotation:

> 'Peace is not freedom from labour – Victory, even when it is complete, is not an end but the beginning of an infinitely harder task. . . After the last war we fell into the error of regarding peace as a state of tranquility in which men took their ease. We are not likely to make that mistake again.'

The Pier and Harbour Commissioners of 1945 gave no indication of wanting to take their ease. During the previous decade they had made Stornoway into one of the best fishing harbours in Scotland.

They had reached the limit of their borrowing powers and committed their successors to repayments which still had over twenty years to run. They had done it all in the belief that the port would never survive economically otherwise. Nevertheless, it had required a huge amount of courage and steely determination to carry it through. And they were not finished yet. Facing an uncertain future, with a kippering trade and a local fleet in decline, and with seemingly no means of funding any improvements, they decided to look at one of the major improvements which had been abandoned in the early 1920s.

The Consulting Engineer's report on the post-war development scheme had been considered at a Special Meeting on 27th February 1945. The six Commissioners present studied it very carefully before, on the motion of Donald J Maciver seconded by Norman Maciver, adopting Scheme 'B' which was divided into four stages totalling £138,000:

(1) 300 ft extension to King Edward's Wharf £67,000

(2) Level down Goat Island and construct embankment roadway £29,000

(3) Construct slipway on Goat Island to take 2 drifters and 1 trawler £35,000

(4) Construct jetty alongside slipway £ 7,000

It was decided that the priority was the extension of King Edward Wharf and they rejected Mr Henderson's alternative suggestion of a 200ft extension which was £19,000 cheaper than the longer one. The reconstruction of No 3 Wharf was deferred for the moment on the grounds that it was in shallower water with rocks in the vicinity.

Archibald Henderson made a convincing case for the improvements, stating:

> 'Stornoway has a fine, natural harbour, sheltered from all winds and with an easy entrance permitting large vessels to reach the shelter of the outer harbour in any state of the tide or weather but, because of insufficiency of depth of water at the quays, there is considerable inconvenience and delay to shipping traffic as vessels can go alongside or leave only at certain states of the tide, and there are no repair facilities whatsoever.'

It was essential, he continued, to have two or three berths available for the accommodation of the larger classes of ship trading at the port including those engaged in the foreign export of herrings. It was also essential, in a port situated far from other ports such as Stornoway was, to have facilities for examination, cleaning, painting and repair of vessels. He suggested a slipway capable of taking three herring drifters at once, as well as providing facilities for the trawlers which frequently used the port. The Commissioners backed this view, Norman Maciver saying they should look not to the past but to the future. The type of slipway recommended by Mr Henderson was of the same type as those at Grimsby, Hull, Fleetwood, Peterhead, Fraserburgh, and Macduff.

Roderick Smith had not been present when these discussions and decisions took place but at the next Harbour Commission meeting on 13th March he congratulated the Commissioners on their foresight and courage in undertaking such a large scheme of improvements. However, he recalled the meeting twenty five years earlier, attended by Lord Leverhulme, when the connection to Goat Island had been discussed. Both the late ex-Provost Maclean and ex-Provost Smith himself had contended that the causeway should be at the Battery end of the Island,

and Lord Leverhulme, after hearing the discussion, had agreed. Norman Maciver said that he had been unhappy with the Goat Island scheme and had telephoned Mr Henderson who explained that his recommendation of siting the causeway at the inner end of the island had been made only on the grounds of cost which would be £50,000.

'The Commissioners unanimously took the view' reported the *Stornoway Gazette*, 'that, if the difference was only in the neighbourhood of £20,000, the scheme for a causeway at the Battery end of the Island was preferable because of the other advantages it offered such as the large area of sheltered water it would provide.'

Mr Henderson, meanwhile, had completed the revised plan of the Goat Island causeway scheme along the lines of Lord Leverhulme's original scheme. He forwarded four sketches with costs which were considered at a Special Meeting on 25th April 1945:

(1) his own original 'Embankment' idea at £29,000;

(2) the 'Embankment' idea on the Leverhulme scheme site, with the roadway widened to 60 ft at £73,000;

(3) the 'Causeway' idea with a concrete wall as proposed by the Leverhulme scheme at £40,500;

(4) the 'Causeway' idea as above but extended to meet the Island at £63,500.

The report stated that the Commissioners had never contemplated undertaking the full development in the Leverhulme plan which had provided for the reclamation of land and the building of quays and which the Consulting Engineer estimated at a total cost of £521,000. However even if they tackled only the provision of a causeway it would cost £73,000 to build this at the outer end of the Island, more than doubling his first suggestion. The Leverhulme causeway, with its straight concrete face on the seaward side would be very wet at times with water being thrown over it, and all that for a cost of £63,500.

The Harbourmaster's view was that they should consider developing the natural inner harbour which was protected from all winds. The provision of a retaining wall on the Castle side would increase vessel accommodation by fifty per cent. Norman Maciver was reluctant to give up the Goat Island plan with its potential site for a slipway, although he admitted that it was 'sound planning to develop a harbour shorewards instead of seawards.'

The very last item in the Minutes 1945 read:

> 'the meeting unanimously decided to make known to the Minister of War Transport its support of the representations made recently by

the Town Council concerning the speed of the proposed new mail boat.'

By that time the war was well and truly over and the world was changed again. While they had made their plans around their meeting table over the months, the United Nations had been established in San Francisco, there was a new Labour Government and a new Prime Minister. Japan had surrendered in August after the bombing of Hiroshima and Nagasaki and a new kind of warfare would face the next generation.

William MacLauchlan

Taking home a 'gad' of herring, 1950s. There was usually a 'fry' available at the quayside from a friendly fisherman when the catches were being landed

Brave New World – after World War II 6

THE FIRST FULL YEAR of peace time in Lewis began with a short but violent storm in early January, the worst gale since the hurricane of 1921. Meeting on 8th January, the Stornoway Pier and Harbour Commissioners had a varied agenda to deal with, a mixture of old problems and new plans.

The aftermath of the military occupation of the harbour, with its disposal of Nissen huts and air raid shelters, and its compensation claims, would remain in the Minutes for another few years.

One unusual item on the table was the Welcome Home and Remembrance Fund which was organised by the Lewis branch of the British Legion and which had as its objectives 'facilities for the well-being and betterment of the community.' The Commissioners delegated Provost Mackenzie and Ex-Provost Smith to represent them at a meeting of this committee which was being widened out to include members of the local public bodies.

The local British Legion was troubled by the circumstances facing the returning ex-Servicemen. The housing situation was desperate. Families were living in overcrowded and sub-standard buildings, with little prospect of better accommodation. In an echo of the 1920s some men had already raided a field at Sandwick and staked out plots for themselves, others were threatening to raid farmland at Laxdale. The Legion's committee believed that 'much could be done immediately to alleviate the distressing housing conditions in the island.'

The Lewis Association also warned the Scottish Office that the island was heading for a crisis of unemployment unless some programme of development was put in place. The Harris Tweed Industry had not yet returned to full production, frustrated by the shortage of yarn supplies for the weavers. Tom Fraser, the Joint Under Secretary of State for

Scotland, arrived on 14th January 1946 for a three-day visit to take stock of the situation and *The Scotsman* commented:

> 'One feature of the Lewis situation which cannot escape Mr Fraser's attention is the queue of unemployed signing on at the Employment Exchange almost directly opposite the hotel where he is staying.'

A New Order

Like the other local public bodies of the day, the Harbour Commissioners were intent on reviving and renewing the economy of the town as far as it lay within their own remit, and their first task was to gain the legal powers to do that. This took the whole of 1946, starting that very first meeting with the consideration of which new provisions should be added in. According to the *Stornoway Gazette*, during a very long discussion 'quite a number of important and knotty problems were raised by the Secretary.'

At the end of the meeting the Commissioners had agreed on the new powers they wished to assume. They wanted to be able to increase the Schedule of Dues by 33¹/₃%. In order to control the crowds meeting the mail, they decided to introduce a penny charge for people passing the barriers. They also asked for powers to levy charges for the use of the proposed Boat Slip, although this was still a long way off. Another provision was for powers to collect dues on general cargo through the shipping companies, to whom they would give a 5% collection fee. Although this would cost the Commission around £140 per annum, they felt it would still be a saving on the clerical assistance costs involved in recovering the small amounts payable. They also wanted to be able to grant leases for the Gifted Properties and to make a charge for parking on their land. Captain Macdonald felt that the parking charge might never be used but the Commissioners should have the powers anyway 'because they could not foresee what might happen in the future.'

An item which took up a considerable amount of time was the question of landing dues on motor cars. They settled on a sliding scale of charges: 5s for cars up to one ton; 7/6 for those between one and two tons; and 15s on cars over two. The item 'Carriages with springs etc' was deleted from the Schedule of Dues at last and replaced by 'Motor vehicles and carriages'.

At the end of the month they held a Special Meeting to go further into the details of new Order, dealing first with specific dues and charges, then moving on to the trickier questions of procedure and suffrage. At the time, the qualification for a vote at a Harbour election was the payment of £2 harbour dues per annum. Roderick Smith's suggestion that this be dropped to £1 was defeated by five to two. It had been moved in his absence by Captain Thomas Macdonald and Bailie JS Maclean who also moved that the Constitution be changed to remove the three members representing the Stornoway Trust and the Sheriff Principal. They wanted to increase the Duespayers representation correspondingly from three to six.

Edwin Aldred and Provost Mackenzie, both Trust members, instantly moved an amendment that the Trust and Sheriff representation should remain the same. They did, however, add the proviso that the Duespayers be allowed to elect another two members. Donald John Macdonald and Norman Maciver moved that there should be no change at all. After 'a lengthy discussion' the eight members present voted 6 to 2 for the *status quo*.

During February and March the Commissioners shaped and honed the new provisions they wanted to add – powers to charge dues on seaplanes. In April they revised the definition of the harbour limits, taking them. . .

> 'from Holm Point on the east following the high water mark to the junction of the seaward face of wall situated south of the Bayhead Embankment thence along the seaward face of the said wall and weir westwards to high water mark on the west side of Bayhead thence southwards along high water mark to Arnish Point on the west and seawards to an imaginary straight line drawn from Arnish Point. . . to Holm Point. . .'

The total costs of the new works were now estimated at £179,000 and the Commissioners decided to make an application to the Treasury for grant for the whole project. They also agreed to 'urge that an allocation be made now of a sum to meet the whole estimated cost of Seventy Five Thousand Pounds for the extension of King Edward Wharf (Work No 1)'.

Malcolm K Macmillan, MP for the Western Isles, gave them his support in a letter to the Secretary of State. No reply had been received by December 1946, although, in Mr Macmillan's words, 'The Secretary of State gave his strongest assurance to me that any practicable means of securing development in Stornoway and Lewis always receive the most sympathetic consideration.'

In the Spring of 1947 the Highlands and Islands Advisory Panel added its support to the Stornoway Harbour development scheme, telling the Secretary of State that it should have priority. The Panel also recommended that the Herring Industry Board should build a processing plant at the harbour to cope with surplus catches.

The Provisional Order finally received Royal Assent on 10th

December 1947. The new Schedule of Dues had also been approved by the Scottish Home Department. Items removed from the 'Exemptions to the Schedule' were goods landed for Lews Castle and the Estate in the days when the proprietor was in residence; and the 'furniture and luggage of fishermen coming to and returning from herring fishings' which, the Minutes said, was a thing of the past. Taxi-cabs and motor cycles were added to the charge for private cars, and although a scale of charges for seaplanes using the harbour was discussed, this was abandoned because of its complexity. The new Schedule was authorised by the Secretary of State to take effect as from 1st April 1948.

Back to Normal

Throughout 1946 life at the harbour slowly re-adjusted to peace time circumstances. In February 1946 the Ministry of War Transport sent out a circular stating that it was proposed to re-start the pre-war broadcasting of navigational warnings by 'wireless telegraphy and radio telephony from Post Office coast stations'.

The Government had proved unresponsive to pleas for the retention of the navigational lights on Goat Island and Holm. The Member of Parliament, Malcolm Macmillan, was unable to persuade the Ministry of Transport to intervene, and the Commissioners finally gave up the fight in June 1946. The Northern Lighthouse Board wrote to say that they were going to carry out the Admiralty's instructions to dismantle the lights, but before doing so asked if the Commissioners would take over and maintain them. However, apart from the maintenance costs, the light on Holm Point was outside the Harbour limits at the time and on the motion of Edwin Aldred seconded by the Chairman, it was agreed to take no further action.

In the summer Donald Murray was demobbed from the Forces and returned home. A notification of 'availability for employment under the Reinstatement in Civil Employment Act 1944' was submitted and he was offered his former job of Collector's Assistant at £4 per week He came back to the Harbour Office on 2nd September 1946, at which the services of the junior assistant were dispensed with, followed two months later by those of the senior assistant. The Ministry of Labour and National Service proposed to establish a Local Employment Committee in Lewis and the Regional Controller asked the Commissioners to appoint someone to it. 'No member was agreeable to let his name go forward,' noted the Secretary.

The Stornoway Fish Selling Company put up a fight to retain their purchase of a small brick building on South Beach Quay. Their reply to the Secretary's letter came before the October 1946 meeting :

> 'Will you be good enough to explain to your Commissioners that the two nissen huts are now occupied by boat owners and have been for the last two years with permission from the Air Ministry, and that they have fishing gear up to the value of £5,000 stored in them. The Commissioners know the difficulty our fishermen are up against for storage as it is impossible to get net storage in or around this town, and to confirm what we say the boatowners and fishermen referred to are prepared to meet your Commissioners if allowed and explain their position. . .'

The Commissioners held to their position of procuring these back but added that they would be prepared to consider granting the fishermen a site for storage in a suitable location. The Air Ministry, who had made removal a condition of sale, enforced the terms of the contract and the Fish Selling Company sold the two nissen huts and the brick building to the Commissioners for £80 in the autumn of 1947. The Commissioners immediately rented out one of the huts to the skippers of the drifters *Daffodil* and *Sweet Pea* for net storage at 5s per week.

Although the evidence of their occupation was gradually disappearing from the quays, the military still kept an interest in the harbour. In October 1946 the Air Ministry Mooring Officer, at No 6 Area, Ayr, received permission from the Commission to lay two seaplane emergency moorings within the harbour, subject to the Harbourmaster's direction.

Brand New Ship

At the end of the war the inadequacies of the mail steamer service had come back into public focus. The *Gazette* had reported in August 1945 that a rival company to David MacBrayne's was considering the running of a passenger link between Stornoway and Ullapool. This unnamed company was said to be interested in road transport and intended to seek parliamentary powers to run a thrice-weekly service to connect with a bus service to Inverness. 'It will be a daylight run in each direction,' said the *Gazette*, 'the sea crossing taking 2½ hours.' The Commissioners had discussed this 'with great interest'.

David MacBrayne's and the Town Council were in testy correspondence over the lack of improvement in the post-war service – What had happened to the plans for a new ship on the route? Why had the service to Mallaig not re-opened? What was being done about travelling conditions on the *Lochness*? Commenting on the responses by the company to these questions Provost Mackenzie declared:

> 'We have heard much about the wonderful freedoms set out in the Atlantic Charter. We in the Islands should demand immediate

freedom from the monopolistic masters of the Minch. We must tear asunder the terrific transport tentacles which are slowly but surely strangling our future prospects.'

The Town Council, accordingly, had been pressurising the Secretary of State for Scotland and the Minister of War Transport for an improvement.

In October 1945 *The Scotsman* had reported that the Ministry had examined 'all the captured German tonnage available' to see if they could find a more suitable ship than the *Hebrides* to take over the run when the *Lochness* went for her annual overhaul but that this had proved impossible. According to the same press report, David MacBrayne had applied to the Ministry of War Transport for a permit to build a new vessel. By the Commissioners' first meeting in 1946 the proposed speed of this new boat was causing concern among the public bodies. They agreed to join forces with the Town Council and wrote to the Minister on the matter.

In the meantime MacBrayne's were granted permission to operate a mobile crane weighing ten tons and capable of lifting up to six tons for loading and unloading containers, on condition that it was mounted on 'solid rubber tyred wheels.' They also requested that when the crane came into operation, the mail steamer's berth should be cordoned off and they laid responsibility for this at the Commissioners' door.

'A modern harbour should be properly regulated and controlled,' said their letter. 'At present there is danger to the general public with the movement of cargo trucks and motor and horse vehicles, but the danger will be greatly increased when a mobile crane, carrying heavy loads, is operating in all directions.'

The Commissioners were slightly irritated. The Chairman remarked that MacBrayne's were making a great fuss over the new crane, exclaiming, 'Surely it's not going to be dashing up and down the wharf. There should be no more danger to the public from this crane than from the derricks that are slinging cargo over the passenger gangway at the present time.' Captain Finlayson said that, if the Commissioners wished, he could rope the crane off but added that he had never seen this done before in 'all his years of experience'.

The new vessel *Loch Seaforth* took up the service on 6th December 1947. She sailed in to the harbour to be met by the sirens of the other vessels in port, flags flying from the Town Hall and a large crowd on the quay. She was officially welcomed by Provost Mackenzie for the Town Council, Norman Maciver for the Harbour Commission, and John Maciver for the District Council. Over 225 feet long and with a draught of 12 ft 9 ins, she had a service speed of 14 to 15 knots and could accommodate 500 first and third class passengers, as well as carry mail, cargo, cars and cattle.

During the afternoon hundreds of people took the opportunity of seeing over the brand new vessel. They came away much impressed.

'By Jove, there's not much third class about this,' said one Councillor, looking around the third class lounge. The observation lounge on the upper deck was modelled on that of the *Queen Mary* and furnished with comfortable settees, upholstered in green and dark red. The wooden panelling was bird's eye maple. Mr H T Leith, the General Manager of David MacBrayne Ltd was able to say proudly, 'The *Loch Seaforth* is a luxurious small liner equal to anything on the British coast.'

After the cramped and unpleasant conditions of the *Loch Ness* she was a welcome improvement to travelling conditions for the islanders and in time the *Loch Seaforth* became one of the best loved of all MacBrayne's vessels on the Stornoway run.

The company expected their new ship to attract more tourists to the island and Mr Leith urged business men and public authorities to provide the additional hotel accommodation necessary to meet this increase. David MacBrayne Ltd, therefore, were startled to find that the harbour dues for the vessel were not to be at the usual compounded rate of £750.

This sum had been agreed for the *Sheila* many years before and had remained during the tenure of the *Loch Ness*. The company sent back the invoices they had received for December 1947, but were informed that the Commissioners had no powers under the new Harbour Order to compound dues. The matter was referred to the Secretary of State for Scotland who agreed that any variation in the rates and dues under the current Schedule of Dues was for him to decide. He was prepared to consider an application from the Commissioners under Section 12 (1) of the 1937 Act for a compounded rate for this class of vessel, assuming that 'regard will be had to the increased tonnage of this vessel and to the views of the operating Company.'

The Finance Committee was given the task of framing a submission to the Secretary of State and in June 1948 proposed that the dues should be based on the current dues for a regular mail and passenger boat less 33 ¹/₃%. On the basis of 313 trips per annum, they reckoned the full dues on the *Loch Seaforth* would be £5,508 16s. With the percentage reduction this would come down to £3,673. The Secretary of State, after consultation with the Ministry of Transport, told Alasdair Macleod:

> 'having regard to the circumstances in which these vessels operate

> and in particular to the regularity and importance of the service provided, a rather lower consolidated rate than suggested by your Commissioners might be appropriate.'

He suggested between 45% and 55% or between £2,000 and £3,000. On the motion of John Morrison, seconded by Bailie Kennedy, the Commissioners opted to charge 55% of the authorised rate, which the Secretary of State approved.

Brand New Town

By the Spring of 1946 the post-war planning of Stornoway, instigated by the Commission almost two years earlier, was gathering momentum. Robert Hurd, the noted architect, had been working on some far-reaching plans and he arrived at the end of March, along with his partner, Mr W Schmidt, to outline these to all three public bodies. Very little of his plans for the harbour appear to have been carried out exactly as he envisaged, if at all, although he was instrumental in saving at least two of Stornoway's oldest buildings from demolition.

Robert Hurd was born in England in 1905, the son of Sir Percy Hurd MP and his wife Hannah Cox from Dundee. He was educated at Marlborough and Cambridge and came to Edinburgh in the early 30s to study at the College of Art. He made his career in Scotland, setting up the partnership of Neil and Hurd, with Norman Neil, in 1932. A strong conservationist, he was a member of the National Trust for Scotland, and a regular writer and broadcaster on architecture. At the time of his association the Stornoway Pier and Harbour Commission he was President of the Saltire Society and was engaged in restoration work on Culzean Castle and Loudon Hall, both in Ayrshire. From 1947 to 1959 he was the Planning Consultant for Lewis, for the Ross and Cromarty County Council. The Dictionary of Scottish Architecture quotes another Edinburgh architect, Alan Reiach on Hurd , 'while always concerned with broader issues affecting Scottish affairs, no cause that he thought worthwhile was too small for his support.'

Robert Hurd met the Harbour Commissioners at a Special Meeting on 29th of March and dealt first with his plans for traffic routes around the Harbour area. He then moved on to his ideas about the warehousing and other buildings, reminding the Commissioners that these were long-term proposals. Most of the existing buildings were in such poor condition that they were likely to be ready for demolition by the time the scheme was implemented, he said. The main interest for the Commissioners was the provision of warehousing and Mr Hurd had, like them, decided on complete redevelopment of the north part of the Quay Street/Esplanade area.

He suggested building new two-storey warehouses along Esplanade Quay with open storage on the ground floors and lock-ups above. These would be built around two courtyards and be served by two moveable loading cranes. The Coastlines company was to get a new one-storey building nearer the quayside to allow free circulation of traffic. More warehousing could be built next to Maritime Buildings on the new quay extension. The only exception to the demolition programme here would be the building he described as 'the old Reform Hall now used as stores and sail loft.' This should be renovated and incorporated into the warehousing since it was, Mr Hurd said, 'one of the few buildings in the area which has any architectural interest.' Its design and proportions were very pleasing

The other building which Mr Hurd felt should be preserved was the Custom House (present day Amity House), which he thought was 'an attractive building architecturally' and which would lend itself to renovation as a Harbourmaster's house eventually. A car park 'for the convenience of persons working in the area' could be constructed at the back of the building

Next to the Custom House, on the corner facing the Fish Market, all the new offices were to be sited and Mr Hurd felt the new Fish Market itself should be 'retained in this part of the harbour for reasons of access and amenity.' He was very complimentary about the plans for the new building, drawn up before the war by Sinclair Macdonald & Son, which he said were very attractive. This should be sited only 10 ft east of its present position where it could be seen from all sides.

Norman Maciver, seconded by Provost Mackenzie, moved the approval of the plans with the exception of the warehousing on the west of King Edward Wharf, which was unnecessary at the moment. The Fish Market was to be the first instalment of the scheme, followed secondly by the erection of the block of offices, and the remainder to be completed 'as may subsequently be determined.' The meeting also appointed Robert Hurd as Architect for each part of the work, with the exception of the Fish Market, the plans for which were already drawn up.

The Commissioners first priority was the construction of the Fish Market and in April 1946 Sinclair Macdonald & Son were asked to take up the abandoned project again. The Ministry of Works approved of a Works Licence for the project but stipulated that 'an alternative form of construction to brickwork and slates will have to be considered.' The Secretary replied that locally produced concrete block walling could be used and that prices were being sought for 'Rubberoid' roofing instead of slates. However, he stressed that the prevailing weather conditions and exposed nature of the site made it wiser to use heavy Welsh or Scottish slate, double nailed.

At the end of January 1947 the Secretary of State for Scotland approved the Commissioners application to have the harbour considered a marine work under Part III of the of the Harbours, Piers and Ferries (Scotland) Act 1937. The Commissioners informed the Ministry of Labour and National Service of their proposed new works and asked the Ministry, 'having regard to the degree of unemployment in this area', to exercise its influence on their application for grant from the Development Fund.

At that point there were 2000 unemployed persons in Lewis, mainly demobilised ex-servicemen. For every three persons in work there were two with no job. The Ministry replied that they had already taken the Harbour scheme up with the Departments concerned, asking for a prompt and sympathetic hearing on employment grounds.

Some Commissioners, however, were having second thoughts about a new Fish Market. The Chairman, Captain Macdonald voiced these when he said it was unwise to use all their £14,000 funds in hand on one building. They would have to charge very high rents to recover the outlay, with no guarantee of being able to let them. He felt that the real priority was the provision of stores, and that the Fish Market reconstruction should be shelved and the existing building re-roofed instead. Norman Maciver argued against this, saying that the Commissioners '. . . should take the long view. If Captain Macdonald's predecessors in the chair had not taken a long view, Stornoway Harbour would not be in the position it was.' This view prevailed at that meeting and the Building Licence application went on.

When the Commissioners met for their Annual General Meeting a month later, on 6th May 1947, the main item of business was the response from the Treasury to their Development Fund application. They found that it had not only been successful, but the Development Commissioners wished it to be carried out as one whole project rather than staged. This was 'in the interests of fisheries and ancillary occupations,' stated the letter. The Scottish Home Department had also advised that a more realistic estimate of the cost would be £212,000, rather than the £179,000 in the application.

The Treasury, therefore, was offering an Advance of £100,000 in grant; a loan of £85,000 repayable over 30 years at 2 ½ %; and an interest free loan of £27,000, not repayable for a period of five years, after which period the Treasury would decide whether repayments should begin, taking account of the Commission's finances at the time. One of the conditions attached to the award was that all materials and all manufactured articles should, as far as practicable, be of United Kingdom origin. Another was that the works should be carried out by direct labour and that the rate of wages should not be higher than that paid to the Commissioners' own workmen. The Commissioners, according to the Minutes, expressed their gratification and accepted the conditions

Ex-Provost Smith, unable to attend the meeting, had written to express his pleasure at the news, which was read at the meeting. He said he was also glad to know that 'aspects of a scheme which I advocated thirty years ago. . . before Lord Leverhulme's plans were heard of has now been approved and Stornoway Harbour will have an enclosed area on the Newton side of Goat Island.'

'Courage Rewarded' said the *Gazette* headline that week, reporting that it was the largest ever grant awarded to the Stornoway Harbour Commissioners. The newspaper added that it was possibly the largest ever public works scheme ever undertaken by a public authority in the Hebrides. Although acknowledging that the Government award was probably made to alleviate the problem of high unemployment on the island, the article gave the Harbour Commissioners much of the credit:

> '. . . it is certain that the scheme would not be going ahead now but for the enterprise of the Stornoway Pier and Harbour Commissioners in having their plans ready, although, at the time they were prepared, there seemed little hope of even an instalment going ahead in the near future, and none at all of the whole scheme, being carried out except over a period of long years.'

The Scotsman added its praise to the Government for its speedy action on the first of the Highland Advisory Panel's recommendations. It, too, noted that help was needed from the Herring Industry Board, including the building of boats for charter by the local fishermen. 'Certainly it would be anomalous,' the article said, 'were the main benefit from the develop scheme to accrue to trawlers from southern ports fishing these northern waters.'

In view of the sudden change in their circumstances, the Commissioners took a fresh look at their priorities. Top of the list was to invite tenders for the extension of No 1 Wharf. The new Fish Market plan was abandoned due to the need for immediate funds to promote the new Harbour Order, and the difficulty of obtaining materials. The existing building would need to be re-roofed and repainted and the tenants were informed that they could stay, although their rents would be raised by 50% from Martinmas. The Commissioners were gearing up for business again.

New Ways of Fishing

The first herring to be sent directly from Stornoway to the continent since the outbreak of war was loaded on the Newcastle registered 'Cyprian Coast' on Friday 97th August 1946. The consignees were Duncan Maciver Ltd and the cargo was bound for the United Nations Relief and Rehabilitation Administration in Hamburg. Although the herring trade would never again flourish to the extent that it had in the early years of the century, considerable efforts to develop it were being made in 1946.

The Herring Industry Board's eleventh Annual Report, published in November, announced their proposals for the freezing of herring and they contacted the Harbour Commission with a query about available quay space.

Alasdair Macleod reported in December on a meeting he had had with Board representatives in Edinburgh, in which they had indicated their intention to 'undertake certain developments at this port.' They were particularly interested, he said, in sites on North Beach and Esplanade Quays, asking for the dimensions of the NAAFI building, the M L Hut and of the two yards on Esplanade and the buildings around and between them. The Secretary referred the Commissioners to the HIB Annual Report which showed 'a proposed plant capacity by 1951 in respect of Stornoway and Ullapool jointly of 60,000 crans'. The Commissioners supplied all the information requested and resolved that 'every possible facility be afforded to the Board to put their proposed developments at this port into effect.'

It would be over a year before the Herring Industry Board was able to put these developments into effect. When the Commissioners received the go-ahead for the new harbour scheme in the summer of 1947, they asked the Herring Industry Board what their intentions were but received no definite response. On 17th September 1947 the Commissioners met the Fisheries Group of the Highlands Advisory Panel, a body set up by the Secretary of State for Scotland to advise him 'on the carrying out of concerted plans by Government Departments, Local Authorities and other public bodies for promoting the best economic use of resources and capacity in the Highlands and Islands.' The Chairman of the Fisheries Group was Alex Anderson MP (who had once taught at the Nicolson Institute) and its eight members included Mrs Naomi Mitchison, Provost A J Mackenzie of Stornoway, Malcolm Macmillan MP, Colonel Henderson , and Sheriff Cameron.

Shooting the Nets

At the end of 1946 the herring shoals themselves presented Captain Finlayson with a problem which ended up in the courts for another two years. The fish provoked a dispute between the local fleet and the Harbour Commission by deciding to come right in to Stornoway harbour and provide an opportunity too good to be missed. A report in *The Scotsman* of 12th December, headed 'Rich Harvest causes Harbour Trouble', described the congestion caused by the drift nets set overnight at the harbour entrance. A southerly gale prevented them being hauled and they were now presenting a danger to shipping. Some boats had made catches worth £250.

Two days earlier Captain Finlayson had reported to the monthly meeting of the Commissioners that, in view of the grave danger presented to shipping, he had posted a general notice prohibiting the setting of nets within the harbour but some fishermen 'although personally approached, had refused to desist.' The Commissioners agreed to prosecute the skippers of the boats concerned.

The herring came back into the harbour a few weeks later in January 1947 and for a whole week hundreds of gannets fed on the shoal 'from dawn to dusk', said the *Gazette*. The paper described the sea one day as white with the splash of the diving birds from the point of No 2 to Goat Island. The fishermen, while admiring the birds' spectacular performance, made the most of the opportunity themselves and set their nets again.

On 31 January five local skippers were charged under the Harbours, Docks and Piers Clauses Act of 1847 with failing to obey an order of the harbourmaster.

When the first case, that of the skipper of the motor fishing boat *Provider* came up on 6th February the Harbourmaster testified how he had had to remove nets from the fairway several times.

Captain Finlayson, reporting the events to the next Commission meeting, said, 'It would appear that the fishermen do not yet seem to realise the serious danger to shipping consequent on filling a commercial harbour such as Stornoway with obstructions of this nature.'

Norman Maciver reminded the meeting, 'As a harbour authority, we are responsible for seeing that the harbour is clear for navigation, and if a a ship gets disabled through foling her propeller with nets, there would be a very heavy claim lodged against the Harbour Board.' The Commissioners decided by five votes to three to seek a new Provisional Order.

The *Gazette* reported that Alasdair Macleod, 'with a merry twinkle in his eye' remarked that the Harbourmaster would never take action against anyone for setting an innocent salmon net at the mouth of the Creed. Councillor Morrison said that the Committee had discussed this but, since one could catch herring with a salmon net, they had not made and exception of it.

Ex-Provost Smith found another potential danger with that suggestion. It was fine at the moment, he said, because the Commissioners knew their Harbourmaster, but they had to consider that, in twenty or thirty years time, these powers might be exercised 'by quite a different person who might enforce the regulations too stringently.'

The Harbourmaster tried to bring the discussion to a some sort of conclusion by saying that if the Commissioners gave the authority to leave salmon nets where they were not dangerous, the difficulty would be solved. Ex-Provost Smith still concerned with poachers, said that if the Harbourmaster knew that nets belonged to salmon poachers, 'he could use his discretion.'

In the end it was agreed the clause should prohibit the setting of all nets other than ground nets and 'nets made specifically for catching salmon.' They also agreed that the fine for contravention was to be £100 or three months imprisonment.

A Tribunal made a decision which approved the Order, with an amendment to one clause, and it began its journey through Parliament, finally receiving the Royal Assent on 16th December 1948.

'After the expense we went to prevent the laying of nets in the harbour, there was no shoal of herring seen in the harbour this season,' remarked the Chairman at their February 1949 meeting. ❑

At their meeting with the Group, Norman Maciver emphasised the need for processing facilities to be built at the port, as had been done in Lerwick. The fishing in Shetland only lasted three months whereas in Stornoway it lasted for nine months of the year, and therefore he felt the Herring Industry Board should provide the same facilities there. This point of view was a popular one, since the winter fishing season had once been the most lucrative of the year. Between the wars German klondyking operations and the Government's pact with Norway to import excess catches of herring to Britain had almost killed it off. It had picked up somewhat during the World War II years but had never come back to its former strength. Now, with modern mechanical methods, some in the industry saw a chance for revival.

Norman Maciver admitted that it was now difficult to get Lewis women to work at the gutting, many having found jobs in the growing Harris Tweed industry. 'Whether they will eventually go back to it or whether herring can eventually be gutted by machinery as efficiently as the women did it, I cannot say.' He believed that the future of the Stornoway industry depended on the modern processing methods.

Bailie Kennedy criticised the destruction of the west coast fishing through Norwegian competition and suggested that the Government

Klondyking

On Saturday 7th February 1948, a fleet of 24 boats arrived in Stornoway with over 1100 crans between them.. Of that, 150 crans were sold for the home market and the balance was taken by the German klondyker *Butt* for the market in Hamburg. However after the sale, a flurry of telegrams arrived with late orders for the home market but these could not be met because they were allocated to the klondyker.

John Bain of Louis Bain Ltd tried to contact the Ministry of Food in London but it was closed until Monday and so the orders could not be met. The Herring Industry Board countered the criticism which arose over this situation by pointing out that the orders had come in too late and that, if the klondyker had not been in port, there would have been a glut of herring on the market. The *Butt* sailed on the 9th with about 2000 crans, the largest ever shipment of fresh herrings from Stornoway. ❑

might consider using some of that country's methods to encourage and regulate the industry. Ex-Bailie Maclean asked, 'if Norway could import fish offal from Stornoway and make margarine and perfume from it, why could Stornoway not have a processing plant of its own?' The Lewis Association wrote to Joseph Westwood, the Secretary of State for Scotland, saying that 'nothing can contribute so effectively to the restoration of the Scottish herring industry as the exploitation of the Stornoway winter fishing season by the establishment of modern processing plant at the port.'

The Ministry of Food released figures of herring landings at Stornoway from January to April 1947 – 6000 tons as opposed to Norwegian imports of 31,000 tons in the same period. However, the *Stornoway Gazette* noted on 27th February 1948 that 'in less than two months 17,747 crans have been landed at Stornoway valued at £70,398 as compared with 9,523 at £41,315.' A new initiative was launched that week when Mr Goodwin, General Manager of the Herring Industry Board, met the Pier and Harbour Commission and the Town Council. It was led by Norman Maciver, who suggested that a new company should be set up to develop the fishing industry. 'If the local population of the island of Lewis – fishermen, fish curers, tweed merchants, grocers, chemists, coal merchants etc – combined with a view to floating a fairly large sized company to develop the most modern ways of handling herring that would be one of the first steps in the right direction, provided they could be assured of the guidance of the Herring Industry Board and the benefit of the experiments which the Board has carried out,' he said. Mr Goodwin welcomed this idea wholeheartedly, saying it would help to end 'the appalling sectionalism which has done so much harm to the industry.'

Norman Maciver echoed the sentiment, saying he regretted the differences between the catching and the processing of the industry. He stressed the financial burden of the new works taken on by the Commissioners and said that they could only meet their commitments if the harbour was used to its fullest extent. Mr Goodwin underlined this point by saying that the people they had wanted to come to Stornoway had been discouraged by the disputes in recent years 'but if harmony could be established between the different sections of the industry,' he believed interest in the port could be regenerated. He outlined the Government's promise of funding towards some of the Board's proposals, but this was limited to £75,000 which 'would not go very far nowadays.' The Board had to look to commercial enterprise to follow up their experiments.

Mr Goodwin also put the Board's plan to a mass meeting of the fishermen, asking them to think it over before coming to a decision. The herring catches were to go to several processes: some for kippering and freshing; some to be exported to America; and the offal was to be reduced to make edible fats, oils and other products. Initially the reduction process would be carried out at Wick but the Board hoped to build a factory in Stornoway in the near future. The fishermen were to be offered a guaranteed price of between 60s and 65s per cran of herring landed. Because the problem of getting gutting crew was worse in Stornoway than anywhere else, the Board would send up six gutting machines which had been trialed at Yarmouth the year before.

Mr Goodwin was at pains to stress the need to face up to the changes in the industry.

> '. . . a large and growing proportion of herring is being handled by machines of all sorts. . . it is very nice to have machinery to do the heavy or dirty work. . . but the use of machinery brings responsibilities as well as benefits. A freezing plant, for example, is not at all flexible. It has a certain daily capacity; you can't just press a button and make it handle more herring just when you want it to do so. The same is true of canning plants. . .'

In the past, the workforce could be taken on and laid off according to the size of the landings, but machines had to be kept going every day if they were to pay their way. Full supplies for the machines could only be got if the herring were there and to restrict their fishing for fear of a glut was counter-productive and 'day to day surpluses which go to meal factories are not the evil you imagine them to be.' He also made the point that the Board could not justify spending heavily on plant and machinery 'unless they can rely on full co-operation and all-out effort on the part of the catchers.'

The fishermen accepted the scheme but, after ten days into the summer fishing, the Stornoway Port Committee was threatening to pull out unless the Board cancelled the licences of ring-net boats fishing out of Stornoway. They claimed the size and quality of the ring-net herring was inferior to the drift-net fish and would lower the prices. Mr Goodwin returned to Stornoway to urge that there was room for both within the scheme but the gathering of 250 fishermen, including many from the East Coast, were unanimous in their opposition. However they agreed to stay with the scheme under protest.

At last, in October 1948, the Pier and Harbour Commissioners received the application from the Board for a site on which to build the proposed plant. They asked for 200 ft by 50 ft site on Esplanade Quay and also for permission 'when loading vessels at Esplanade Quay, to set up a mechanical conveyor across the space between the front of the proposed factory building and the edge of the quay, as speedy handling from the cold store to the ship's hold is essential if the full benefits of freezing are to be gained.'

Ships and Aeroplanes

• The summer of 1948 saw two visits from small vessels fleeing the new regime in their home countries. In mid-July a small motor vessel entered the harbour carrying a party of about 30 refugees from Estonia. The *Astrid* was 13 registered tons, 47 ft long, and carried sails in addition to her engine. She had been bought collectively by the families on board, who had all fled to Sweden when the Russians took over their small nation. Some of the men were seamen and fishermen, others were farmers, but they all came together while working in Swedish factories. Still uneasy about the closeness of their Communist neighbour, they decided to seek a new life in North America. Between them they saved enough money to buy the *Astrid* and equip her with fuel and provisions. Then, with their wives and children, they sailed from Gothenburg on 5th July, bound for Canada, where they had friends.

For the first four or five days the journey was unpleasant; the women and children were all seasick, the weather deteriorated, and the engine broke down. For a day and a half, the *Astrid's* skipper, Herman Vilu, continued under sail, before altering course for Stornoway to shelter until the weather improved. The vessel arrived in the harbour on Sunday 11th July 1948 and became a worldwide centre of attention at the quay for the rest of the week. The population of Stornoway took this brave little party to their hearts and sent aboard fresh milk, fish and chocolates, to supplement their spartan provisions of salt fish, salt pork and biscuits. Captain Vilu thanked the townspeople for their kindness.

Doubts were held by many as to whether the little boat would make landfall, but make it she did. At 4 am on Saturday 17th July she sailed from Stornoway harbour. At she reached the open sea, she met the Stornoway fishing fleet returning one after the other in line. The first drifter passing the *Astrid* sounded her siren and the Estonians responded by waving frantically. The *Gazette* described the early-morning scene – 'As she sailed past the line of fishing boats, each saluting in turn, it was almost like the Royal Barge taking the salute at Spithead.' Exactly fourteen days later, on Saturday 31st July 1948, she sailed into St John's Harbour, Newfoundland, and Captain Vilu sent the agreed telegram to Provost Mackenzie – '*Astrid* arrived St John's 31st 7th O.K. Capt. Vilu'

• Apart from ships and herring nets, Captain Finlayson found himself in trouble over an aeroplane in 1949 – several large pieces of aeroplane to be precise. The request by British Bitumen for extra space on South Beach Quay had prompted the Commissioners to repossess Station No 62 there from the tenant, Mr Percy Marvin. They gave him one month to remove the very large accumulation of scrap metal which had been dumped there. Mr Marvin replied that he was at a loss to know what to do with this very unwieldy material. 'I wish it to be clearly understood that I am not the owner of this material although the station is let to me,' he wrote. He did remove one or two lorry loads of the smaller pieces but the bulk of the metal remained in situ. Running out of patience the Harbourmaster took over and set about the removal himself. He hired a lorry and trailer and headed with the load for Marybank, and the site used by Messrs Pink and Macritchie, who were alleged to be the original owners of the scrap.

The operation went well until the lorry turned into the side road at Marybank when the trailer sank up to its axles in the peat and had to be hauled out by the RAF crane. However his troubles were only just beginning because, although the trailer was back on terra firma, it was still pointing south and filled most of the width of the main road. There was no chance whatsoever that it could be turned and taken back to South Beach Quay, so the journey continued, accompanied by several policemen regulating the traffic, in the hope that another side road could take the weight or allow a turning. It was a vain hope and, as the Harbourmaster reported to the Commissioners, 'I was reluctantly compelled to dump this and several other large pieces of the same material at the road side near Loch Lathamul on the Lochs road where there was already a considerable accumulation of rubbish which is now more or less covered by this latest addition of clean material.'

Confident that he had found the answer, Captain Finlayson despatched a second load, complete with police escort. The last large piece was already loaded on a trailer for transport to Lochs when the County Sanitary Inspector got on the case and prohibited the dumping of any more scrap at the roadside. This the Harbourmaster disposed of behind the old slipway wall on South Beach 'where it is out of sight and out of the way meantime,' he said.

Provost Mackenzie said this explained why he had recently been asked if a plane had crashed near the Lochs Road. He said it was 'screaming scandal that scrap should have been dumped at one of the prettiest lochs in Lewis and the people responsible should be rapped over the knuckles.'

'You are one of the people responsible,' replied the Chairman. 'You are a member of the Board and responsible for the actions of your employees.' Provost Mackenzie said he hadn't been party to any such instructions being given to anyone. The Harbourmaster said there had never been any intention to dump the scrap at Loch Lathamal but they could not turn the trailer, adding that the simplest way to dispose of it would have been to dump it at sea but since there was always the danger that someone might claim it and ask them 'to fish it up.'

Both the County Council and the Stornoway Trustees were insisting on the Commissioners taking steps to remove the aeroplane scrap on the Lochs Road. The Commissioners replied that they were anxious to do so but that it was proving a problem. They also launched a counter attack by asking the County Council if they could suggest a suitable area for dumping this kind of material 'as it is evident from the number of dumps containing various kinds of scrap metal situated in numerous places in proximity to rural roadways that the provision of such a place is indicated.'

This cut no ice with either body, both of whom wrote back to the December 1949 meeting asking what steps were being taken. The Commissioners continued the matter, informing both that it 'will not be lost sight of but will be undertaken in due course when the problem of disposing of it satisfactorily has been solved.' The solution would not be found for another eighteen months. ❑

Western Isles Libraries

The little Estonian vessel *Astrid* alongside in Stornoway before making her two-week journey across the Atlantic in July 1948

Herring girls 1950s – Carrying the baskets are Dolina Mackay, Upper Bayble; Mrs Bremner, Fraserburgh, and Shonag Finlay, Calbost & Manor Park. In the background are Magdalene Macaskill, Sheshader & Marybank; Annie Mackenzie, Back & Manor Park; and Mairi 'Bakey' Macleod, Coulregrein. Annie Mackenzie's son became Harbourmaster in 1978

Norman Mackenzie

J Allan Cash, London

Fishing for cuddies at King Edward steps – a favourite pastime for generations of Stornoway 'wee coves'

William MacLauchlan

William Watson

(Opposite)
A peaceful Inner Harbour 1950s – among the boats alongside are *Cynosure* SY483; *Home Rule* SY514 and *Agnes Irvine* SY153

Norman Mackenzie

Salvaging a little salt, 1950s. The harbour was a magnet for previous generations of Stornoway children, who romped among the curers' herring barrels and played hide and seek around the bales of wool landed for the Harris Tweed mills

A stroll with grandpa, North Beach 1960s

Norman Mackenzie

Sandy Matheson

MacBrayne's 'small liner' *Loch Seaforth*, arriving on 6th December 1947
to take up the mail run

Stornoway Gazette

James Shaw Grant
The historical record of the Commission and the island owes much to the journalistic talents of James Shaw Grant and his father

Pacific Coast, only one of the many Coast Lines weekly cargo boats from 1925 onwards. She did the last run in November 1964

The Royal Yacht *Britannia* dressed overall. She visited Stornoway twice in her career and frequently sailed around the islands during the Royal Family's summer breaks

The *Lochness*, regular mail steamer from 1929 until her replacement by the *Loch Seaforth*

3 photos – Malcolm Macdonald

Stornoway Pier and Harbour Commission.

Rates Payable under the Harbour Piers and Ferries (Scotland) Act 1937
Incorporated in the Stornoway Harbour Order 1926 and 1947.

Inclusive of Drinking Water (5s)

FISHING SEASONS—

Summer—From 1st May to 31st July.
Autumn—From 1st August to 11th November.
Winter—From 12th November to 30th April.

	Season's Dues	Trip Dues.	
		First Trip	Thereafter
Steam Drifters, Motor Boats and Sailing Boats 40 feet Keel and above	76s 2d.	10s 10d.	9s 4d.
Under 40 feet keel.	40s 7d.	6s 9d.	4s 10d.

BOILER WATER—

Each time supplied to Boiler	5s.
Each time supplied to Boiler Feed Tank	5s.

Fresh Herrings Landed	5d per £ value.
White Fish landed	5d per £ value.
Liners and Trawlers	4d per registered tonnage.
Vessels wintering or laying up alongside Quay	8s 10d per month.

The slipway at Goat Island under construction 1953-54

The Admiralty water carrier *Freshpool* carrying out the slipway trials, 13th June 1955

A New Era 1950–1969 7

AS THE 1950s dawned the Stornoway herring trade, in spite of all the efforts to revive it, was still struggling and by the middle of January the port, at the request of the local fishermen, had been placed outside the Herring Industry Board's Minch Winter Fishing Scheme.

This situation, according to *The Scotsman* had come about by 'a concatenation of unfortunate circumstances', beginning in mid-December with the cancellation of the German klondyking contract. With the local market now able to handle only 200 crans per day, a 'strike' by the kippering firms brought the herring trade at the port almost to a standstill. The buyers maintained they were seriously disadvantaged compared to mainland firms by the 85s 10d per cran they had to pay as against the 72s per cran minimum price that mainland fishermen were settling for. Their production costs were also higher by around 54s per ton for the necessary imported wood and boxes.

While the Stornoway dispute was going on, the Chairman of the Herring Industry Board, Sir Frederick Bell, was touring the United States promoting the Scottish kipper, taking with him '300 boxes of selected quick-frozen Scottish summer kippers for distribution to interested persons.' He noted that American kippers were on sale in every city he visited but were almost always presented with a touch of tartan or some other Scottish hint in the design of the packaging. Sir Frederick came back convinced that there was a market to be developed in the west and south of the States.

On his return home, however, he found the Board and Stornoway buyers still in deadlock and the East Coast boats heading for the port on 9th January were advised to land their catches at mainland ports. On the following day Alex Summers, the Stornoway buyer for John Woodger & Sons, London, was instructed by his bosses to buy 60 crans at 76s per cran, the Herring Industry Board's recommended minimum price. Over in Ullapool and Mallaig the catches went for 89s 10d. Kenneth Maciver and John Murdo Bain told the press that their firms had now paid off the last of their workers. Bain added, 'We are taking all the risks.' By mid-January there was virtually no market at Stornoway harbour other than catches landed for fish oil and meal. Against this background an advert appeared in the press for tenders for the building of the Herring Industry Board's Ice Plant at Stornoway.

'A Venture of Faith. . .'

For the Pier and Harbour Commissioners, well on their way through a highly expensive improvement programme, this was not an encouraging sign for the future. However, at their first meeting of the decade, they were preparing for the summer fishing season as best they could.

The first ship to berth on the new portion of King Edward Wharf was the destroyer HMS *Wilton* on 18th May 1950, while the first cargo ship to discharge there was SS *Tuscan* of Glasgow, a week later, landing a shipment of road bitumen in drums. On Kenneth Maclennan's suggestion, this information was recorded in the Minute of the June meeting.

Not in the Minute, however, was the fact that the first cargo shipped had been a bag of potatoes taken aboard a small Cromore boat some months earlier. The *Gazette* reported that small detail when it welcomed the completion of the extension, noting that 'many a much more important fishing centre than Stornoway would give a king's ransom to have such a wharf in such a harbour at their disposal.'

The paper also observed that the wharf had come into use without any ceremony or speeches. 'This is in keeping with the tradition of the Harbour Commissioners who have completed the construction of all their wharves in a succession of ambitious schemes without the slightest fuss or show.'

By this time, however, the costs of the latest ambitious scheme were giving cause for concern. The Consulting Engineers were estimating a total cost of £338,000, caused mainly by higher transport costs and possible wage increases. 'The way money is inflating at the present time,' said Donald John Maciver, 'we would be better to spend it now and pay it back when money is cheaper in years to come.'

'When are the years to come?' asked John Morrison, who also happened to be the Town Council's Burgh Treasurer and had earlier in the meeting advocated the Commissioners cutting their coat according to their cloth.

Million Pound Harbour

The phrase 'million-pound' harbour arose from a speech by Norman Maciver at an Economic Conference in the Town Hall. The idea for the economic conference held in Stornoway Town Hall on 23rd February 1951 came up at the Town Council meeting on 8th January that year in view of the serious economic situation in the island consequent on the troubles of the herring fishing and Harris Tweed trade. Pre-war markets for fish were unlikely to return and the rising cost of wool had already led to pay-offs in the spinning mills. The phasing out of the Government coal subsidy by the end of the year would mean higher freight costs for both industries. The council decided that 'the time has come to bring before the Ministers and Government Departments concerned the seriousness of the position.' A Joint Committee was set up consisting of twelve members from various bodies, including Norman Maciver and Edwin Aldred from the Pier and Harbour Commission.

The result was the one day conference, attended by representatives of the Herring Industry Board, the White Fish Authority, the Fisheries Department of the Scottish Home Department, the Board of Trade, the Ministry of National Insurance, the Wool Marketing Board and the Department of Agriculture. From the local side, besides Councillors and officials, there were representatives of the fish curers, the Harris Tweed mills, and the TGWU, as well as members of the Lewis Association.

Norman Maciver summed up for them at the end of the final session of the nine-hour talks, hoping that there might be some far-reaching effects from the frank discussions they had had that day. Speaking of the new harbour works he said that when completed the whole harbour works 'at present day costs, will not fall far short, if it does at all, of £1,000,000 sterling.' He added that this was 'quite a lot of money' and that the Commission, although recent trends might cause some misgiving, were not dismayed. ❑

Rumours that the Herring Industry Board were intending to centralise the Minch fishing in one port, not yet specified, worried Robert Hurd, by now County Council Planning Consultant for Lewis. The harbour improvement works were a strong point in Stornoway's favour, providing facilities which Mr Hurd said were far ahead of the other ports and in this the Pier and Harbour Commissioners could take some comfort, despite the escalating costs of their scheme. The main driving force behind it all, Norman Maciver, returned to the Chair in July 1950 after a long illness and received a warm welcome. He had been elected Chairman in his absence, for which he thanked the members, although, he said, he intended to stand down the following April when he would have completed 30 years on the Commission and by which time the Harbour Scheme would be completed. Mr Maciver was wrong on both counts. He remained Chairman for a further decade, until his death in June 1960; and the Slipway scheme was not finished until the mid-fifties.

'The most desirable sites in the Burgh'

The Commission still held a stock of ageing and dilapidated property which needed continual attention, a new stove here, a sink there, re-roofing somewhere else. The houses, built by the shipowners and merchants of a by-gone day, were now warrens of closes, with families living above fish-mongers, butchers, plumbers and general shops. They had been old when Lord Leverhulme gifted them thirty years earlier; now they were verging on slums. When one of the tenants complained in December 1950 that he had no electric light in one of his bedrooms and that he had to share a meter with another tenant, he was told that 'electric lighting in the Commission's property is entirely a matter for the tenants concerned.'

The post-war programmes of housing and development allowed the Commission to dispose of some of these properties, always with an eye to the best deal. This was seldom a very high return but every penny helped in their current financial situation. One of the best opportunities presented itself in May 1951 when the County Council inquired if the Commissioners would be prepared to sell the property on James Street/ South Beach as a site for County Council Offices. This block had originally been considered as the location of a Community Centre as part of Robert Hurd's town plan, and the Town Council also owned part of the site.

The result of the discussions was the adoption of the Town Council's suggestion to develop the site westwards rather than eastwards, bringing the front elevation of the new building onto South Beach. This would mean the demolition of a block of unfit houses on a main front street. It also meant that the Commission was now the only proprietor involved in the negotiations. By the summer of 1952 the matter was receiving the attention of the District Valuer who estimated the value 'of the owners' interests under burden of tenancy' to be £425, describing the tenement as having 'four houses of two rooms, all let; each two tenants share a WC and there is cold water only. The building is very old and in a poor state of repair.'

The Commissioners refused this offer, on the grounds that 'no regard could have been given to the desirable location of the site situated as it is with considerable length frontage along two main streets close to the centre of and within the business section of the town.' They asked for £1500, telling Mr Ross, the County Clerk that they did so 'with the conviction that the property could easily realise a much higher sum and . . . as a concession to encourage the much needed provision of County Council offices in what is probably the most desirable site in the Burgh for such a purpose.'

The revised figure of £1000 came before the April meeting. There was a difference of opinion among the Commissioners on whether or not to accept this. The Chairman and Bailie Bain were inclined to hold out for the full price, seeing the pace of the County Council responses as 'a half hearted attempt' to acquire the site. Roderick Smith, on the other hand, felt that the Commissioners would be making an excellent gesture if they modified their attitude and accepted the offer. Norman Maciver wanted to test the open market. The ex-Provost pointed out that the properties had been available for years without any interest. 'It was well known,' he said, 'that if any person were prepared to pay a big price the Pier and Harbour Commission would jump at it.'

Councillor Langley produced the compromise that ended the discussion – to advertise the property for four weeks on the open market and, failing a more advantageous offer, to accept the District Valuer's valuation. There was no other offer and the deal went ahead.

Television came to Lewis in September 1959 when the Stornoway firm of Maciver and Dart decided to introduce a TV relay system from a receiver on Gallows Hill above the harbour, to a distribution point at their premises on Cromwell Street. This was to be achieved by running a single co-axial cable from Gallows Hill to the water's edge, then under water to the beacon at Poll a Choire on the western shore of the harbour and on across the harbour to the cattle pens at King Edward Wharf. The cable would then run along the side of the pier to the end of the Coast Lines store, and on from the ridge of this store to the roof of the Fish Market and across South Beach to the chimney head above the Harbour Collector's Office. From then on it made its way up Point Street from building to building. The result was a nightly gathering in Cromwell Street as crowds flocked to watch the latest output from the BBC on two large screens in Maciver & Dart's window, one of the first programmes being coverage of the General Election of 1959.

The old ways were not yet gone, however. A photograph on the front page of the *Stornoway Gazette* in September 1959 showed a sheep shearing fank being held on Cromwell Street quay, after the boat bringing the animals from the summer grazing on the island of Rona had to make for Stornoway rather than Ness. The street wall was lined by interested watchers and a happy Lewis collie stood in the foreground as the shearers worked swiftly on the grass stretch of this pier.

The harbour was the hub of the town for many of the island's travelling public, since extensive use was made of the quay space by vehicles looking for a parking space. Commissioner John M Bain took up, in May 1951, the subject of 'the use which is being made of harbour property by public service vehicles for parking, the free use of the harbour's water supply by lorries for washing and the parking of vehicles at 'Billingsgate' for the sale of fish in respect of which no payment is made to the Commission.' Both Councillor Bain and the Chairman, Norman Maciver, were frequently prevented from accessing their curing stations by the indiscriminate parking of vehicles on the pier. 'I feel very sore about it,' said Norman Maciver, 'I am paying rent for a station there and can't get near it.'

The Commissioners were also aware that certain parties were making free with the harbour's water supply. 'You know that all buses carry a length of hose which they fit to our taps on the pier,' said the Chairman. Captain Finlayson had done what he could to curtail this, turning the water off at some taps except at certain times to allow boats to take supplies.

A Harbourmaster's Lot. . .

Through the decade Captain Finlayson continued his dignified reign over the quays and the changing face of Stornoway harbour as it moved into the second half of the twentieth century. The increased cost of living showed in the last item in the December 1950 Minutes, when the Harbourmaster submitted a letter stating that the increase to £15 of his uniform allowance last given three years earlier 'has not proved adequate to meet the cost of a blue serge uniform suit, which at the present time costs, at the most favourable terms, £19: exclusive of cap and badge.' The Commissioners upped the allowance to £25 per annum.

The Harbourmaster's challenges were many and varied. Captain Finlayson, like all his predecessors and successors, had periodic purges on derelict craft in the harbour. In a report to the Commission on the subject in May 1954 he drew attention to two steam drifters that were not fishing but were being used by their owners for the barking of nets. The Commissioners instructed him to have these vessels anchor in the bay during the fishing seasons to avoid congestion. The Stornoway Fish Selling Company complied with this order but Messrs Bain & Morrison resisted. Captain Finlayson reported that this caused great inconvenience 'thus obstructing other vessels from discharging their fish or otherwise disturbing those which are already berthed and want to be left in peace especially at weekends.'

In September 1953, a letter was received from Mr Main, of the Stornoway branch of the Herring Producers' Association complaining of 'the serious inconvenience caused to fishermen and occupiers of offices in the Fish Market by the continued use of the Market by dockers and other unauthorised persons for the garaging of bicycles.' Mr Main also stated that considerable annoyance was caused to fishermen by 'the daily use of the Fish Market by dockers for gambling purposes, i.e. the playing of cards, pitch and toss etc, without any apparent attempt by the authorities to prevent or suppress the practice.' He added that a considerable amount of serious pilfering was going on and requested that the Harbour Commission take proper control of the Fish Market.

The Harbourmaster came down on the side of the dockers. He said there was no reason why the old established privilege of using the Fish Market should be prohibited as long as it was not abused. It was the only sheltered place in the harbour and always used for a short period before 8 am when the men registered for unemployment pay with an official from the Ministry of Labour; then again in the evening before the mail boat arrived. He declared that in neither case did the dockers interfere with the business of the fishermen and dismissed the claim about pilfering as a 'frivolous complaint'. As for the gambling, there had been one or two occasions during the slack season when men had been found playing cards but this had been effectively stopped. Bicycles were sometimes left in the Market but were not usually left in the way of the users of the building. Faced with this spirited defence, the Commissioners took no action.

By far the biggest event for the Harbourmaster was the visit of Her Majesty Queen Elizabeth, on the Royal Yacht on 18th August 1956. The Commissioners made extensive arrangements for the event, setting up a Committee to handle it all and ten days before the visit further instructions were issued – posters to go up around the harbour prohibiting empty containers left on the pier until after the Royal Visit; additional men to be engaged to assist harbour staff in making sure no 'empties' were deposited on the quays; public notices to be issued requesting boats to refrain from going too close to *Britannia* and restricting her berthing. A motor launch was to be hired for assisting in the supervision of craft and the Harbourmaster was to enlist the help of any naval vessel escorting the Yacht. Permission was given to the Old Folk's Club to set out chairs in the Fishworker's Rest House on the day.

On the day, crowds packed South Beach and on No 2 Pier the spectators were so close together and so near the edge of quay that the Harbourmaster and other officials feared for their safety. Cromwell Street Quay was set aside for the use of buses which brought people in from all corners of the island. In a row of seats along the south side of Maritime Buildings sat members of the Commission and the Town Council, 'who are not appearing elsewhere in the proceedings'. With them were their wives and those of the officers of the Guard of Honour who were waiting on No 1 Pier. A 'triumphal arch' of heather and greenery was constructed by the local florist, George Newhall, between the edge of the wharf and the corner of Maritime Buildings. It carried the welcome *Failte do Leodhas* painted by signwriter Alex Macleod.

Captain Finlayson was on the quay when *Britannia* slipped into the berth on the west side of King Edward Wharf, at the spot where the Queen's great-grandfather had come ashore 54 years earlier. As the crowds waited for the party to appear, the Royal children were seen running about the decks. Princess Anne attempted to climb over the side before being retrieved by one of the crew. She and Prince Charles then watched with interest as the Pipe Band struck below before their parents and their aunt, Princess Margaret, descended the gangway to the red carpet hired for occasion. After the usual presentations and bouquet, the Duke of Edinburgh drove himself and the Queen in a open-topped Hillman, off the pier and round to the Sailor's Home on North Beach Quay, where they were shown round the building.

Prince Philip then drove slowly through the crowds round to the 'narrows' of Cromwell Street and across to the curing stations on South Beach. Norman Maciver conducted the party about the station, explaining the process, as the Queen watched the flashing fingers of the gutters and the Duke admired the swift, neat packing of the barrels. They were watching a dying art. Only two curing stations were working on the Stornoway quays in the summer of 1956 and Norman Maciver's family firm would soon send their *Windfall*, the last of the port's steam drifters, for scrap.

The Harbour officials had a few hours to relax while the party continued their tour of the island. After visiting the Castle College, followed by a black house and new 'white house' in Point, and finally the Tiumpan Head Lighthouse, it was back to King Edward Wharf where Councillor Mrs Urquhart presented the Queen with a memento of her visit, a gold paper knife in the shape of the herring girl's *cutag*. The visitors were piped aboard *Britannia* and all of them with the exception of Princess Margaret appeared on the upper deck to wave to the crowds while the yacht backed away from the pier. The Lovat Scouts played 'Happy We've been a' Thegither' and the spectators remained until the ship disappeared behind Goat Island. It had been a good day.

The problem of scrap on the quays continued to haunt the Harbourmaster with a growing collection on Cromwell Street Quay awaiting shipment. Councillor Langley's suggestion of screening it by planting bushes was met with scorn from Norman Maciver. 'I would rather see the pier crowded with barrels and herring and salt and women as it was years ago,' he said. 'Over my dead body will rhododendrons be planted on Cromwell Street Quay.'

Change at the Helm

The very first year of the 1960s was marked by the disappearance from the harbour scene of two of its longest serving and influential characters. In January the Harbourmaster, Alexander Finlayson, tendered his resignation after 36 years in the post, while six months later Norman Maciver died after a short illness. For almost 40 years, both men had been inextricably linked in the minds of islanders with the running of Stornoway Harbour. It was the end of another era.

Norman Maciver had been a member of the Stornoway Pier & Harbour Commission for 38 years, most of these as Chairman. He had held a strong grip of its business and without his steely determination the infrastructure at Stornoway Harbour would not have been improved so drastically and so effectively. At the Commissioners' meeting, just after his death, the newly elected Chairman, Edwin Aldred, paid tribute to Mr Maciver's 'zealous interest in the welfare and development of the harbour.' A tribute in the *Gazette* spoke of his leaving a permanent memorial in his native town, citing the development of both the harbour and the Lewis Hospital, as well as the introduction of electric light, as his major achievements. . .

> 'The harbour, the hospital and the electric light which burns in almost every home in the island, we owe in some degree to his foresight, his efficiency in planning down to the minutest detail and his great skill in negotiation.'

Captain Finlayson was now in his mid-seventies and his resignation letter explained his feelings:

'Although I have been blessed with almost perfect health throughout my life, I think I would be tempting providence if I carried on much longer as Harbourmaster of this port, for in the words of the famous poet 'no man can tether time or tide' and I have now come to the conclusion that time is fast overtaking me. I would therefore request that I be relieved of my post as soon as this is found convenient. I would like to record my appreciation of the kindness and courtesy shown to me at all times by the Harbour Commissioners, and I am truly thankful that fate has placed me in my old age where I have been happy and contented over a long period of years.'

The post was advertised in the press and a short leet of three candidates chosen. On 1st June 1960, the new Harbourmaster, Captain Alexander Mackay, from Cluer in Harris, took up his duties at a salary of £865 per anum plus uniform allowance and free accommodation in a new harbour house still under construction. Captain Finlayson stayed on for six weeks to break him in gently.

Rebuilding the Finances

For the Commissioners who steered the harbour authority through the 1960s, this decade was a time of rebuilding, both in their financial situation and in their property. Financially, the Commission did well in the last year of the 1950s, earning the highest revenue in their record to the end of their financial year in 1959. Over half of the £27,000 income came from general merchandise, where the main increase lay, although the klondyking operation that winter also helped. However the loans for the extensive harbour improvements began to fall due again as the moratoria allowed by the Government in the leaner years came to an end.

In August 1962 the question of repayment of the free loan of £50,000 authorised in 1951, came before the Commissioners again. The Secretary of State had already granted a five year moratorium on the repayments from August 1957 which was now due to expire.

The Commissioners bought another two years' grace by pleading that the high cost of the harbour improvements had been beyond their control and that slipway and causeway, which although providing an essential service to the fishing industry, had not so far provided enough income to even meet the running costs. When the two-year extension expired in August 1964, the Commissioners tried again to have this sum converted into a grant. The Departure of Agriculture and Fisheries insisted that the loan had to be repaid. They were prepared to give one more year's grace, by which time the Public Works Loan Board loans would be repaid.

'It is suggested,' said the Department in their letter of 22nd October, 'that the Commission should take a decision in the near future on the long-deferred question of increasing the harbour dues and that they might also consider whether it would be possible to effect any economies in their expenditure, for instance in their programme of building works.'

The Department of Agriculture and Fisheries for Scotland continually pressed the Commissioners to increase the Harbour Dues, pointing out that the present Schedule had been in operation since 1956. The Commissioners, however, remained reluctant to do this, 'having regard to the burden of transport costs' on the community, and continually deferred action.

In spite of the fact that the Harbour Revenue for the 1963 financial year was up by £3,763, Alasdair Macleod reminded the members in May 1964 that it had been an exceptional year, due largely to the traffic bringing in materials and equipment for the new NATO base. There would be a drop in the current year as a result of the closure of the Coast Lines service, he warned, and apart from this, it was to be hoped that the new Harris ferry would not have too much effect on traffic at the port. The Department of Agriculture and Fisheries had recommended an increase of 5% in the harbour dues, he said, and although the Commissioners had been reluctant to do this, it would certainly have to be looked at. The dues on prawns were not realistic at all, for instance.

Finally on 19th November 1965 the Commissioners held a special meeting to consider the increase of harbour dues and also to look into the question of levying harbour dues on passengers for the first time. A letter from David MacBrayne Ltd stated that arrangements were being made with all West Coast harbour authorities for incorporating pier dues on passengers into steamer fares. The standard agreement was for sixpence per adult and threepence for children aged 3 to 14 years old. Children under three were not charged. The Commissioners, after discussion, decided to accept this method for the future, since they had never before had powers to levy passenger dues, possibly because of the difficulty in collection.

Other changes to the Schedule were suggested to the meeting and Donald Murray, the Collector of Dues, estimated that the revision would yield, based on current traffic, an annual increase of revenue at £700 to £1,000 on goods; £400 on shipping and £200 from fishing vessels. Using the previous years landings, he calculated the revenue from prawns, lobsters and crabs, at 5d in the £1, to be £944 14s 2d. The meeting reduced the rate to 3d in £1 and set the charge for processed or frozen shellfish to 8s 4d per ton. Several items dropped from the Schedule as superfluous included blubber, cobles, hats and saddlery items. They were replaced by anti-freeze mixtures, electrical fittings and batteries, hardboard, asbestos, and trees.

Dean of Guild James Maclennan had some strong views on the state of the Commission's finances and, in March 1968, he gave notice of motion to be considered at a special meeting. He moved that 'steps be taken immediately to raise sufficient capital to put the whole undertaking on a proper economic footing.' He advocated that these should include charging for car parking on harbour property, appointing a car park attendant, and re-siting the Cromwell Street bus station. Mr Maclennan wanted the Commissioners to consider 'the feasibility of raising capital to purchase plant and equipment to enable the Harbour Commission to make better use of their very extensive resources.'

Mr Maclennan was also frustrated by the Commission's refusal to allow the Lewis Swimming Pool committee to erect a building on Cromwell Street Quay for use as a recreational centre for tourists and youngsters. He thought they were completely out of touch with the needs of the community. The Swimming Pool committee were allowed to run an annual regatta however, complete with greasy pole and boat races. Other clubs springing up around the harbour in the 1960s were the Stornoway Sea Angling Club who were granted a site on South Beach Quay for a clubhouse, Stornoway Sailing Club given a station for a dinghy park, and the Water Ski Club which chalked up thirty members in a month for this new sport.

Rebuilding the Fish Mart

The building which caused the longest and most controversial debate before it was finally built was the new Fish Mart on North Beach Quay. In May 1962 the Department of Agriculture and Fisheries for Scotland made a fresh approach to the Commissioners on the subject of providing a market for white fish. Although the Commissioners were willing to provide a site for such a building, they steadfastly held out for several years against placing it on the quay edge, pleading amenity, and traffic considerations as their reasons. The fishermen and the Department did not give in, the latter repeating its earlier warning to the Commission that the Secretary of State could not offer grant-aid unless he were satisfied that 'such money would be well spent in the advancement of the fishery interest.'

The Commissioners tried again to persuade the Department to accept their second suggestion but to no avail. A letter to the Secretary dated 29th June 1962 gave the reasons:

> 'As you know, a market at the edge of the quay would enable fish to

> be swung straight from the boats into the market and thus eliminate the double-handling of catches and thereby decrease the possibility of damage to the fish as well as eliminating labour costs. In the circumstances you will appreciate that siting the market some 24ft back from the edge of the quay, as now proposed by your Commissioners, while it would speed up the handling of the fish, would not eliminate the double-handling of the catch and to that extent the site now proposed cannot be regarded as ideal. In addition a market on the site now suggested would restrict the available roadway considerably, and the continuation of the practice of barrowing fish across the roadway is not particularly desirable either from the hygiene or the safety point of view.'

At the end of March 1963, a joint meeting of the seine net fishermen of Lewis and the Management Committee of Lewis Crofters Ltd voiced 'considerable dissatisfaction' with conditions at the port. The Secretary of Lewis Crofters Ltd wrote to express this, particularly with regard to the Commission's letting of a North Beach pier site to the Council for a car park. 'It was considered that this decision did not envisage long term development of Stornoway as a fishing port,' the letter said. 'It was hoped this would be the Commission's main objective rather than tourism.' The joint meeting had requested the Harbour Commission to establish an enclosed fish market at a suitable site.

The Chairman, Edwin Aldred, was indignant. 'It is not fair to accuse us of doing something to the detriment of the fishing industry. Bring the fish into the Harbour and the Commission will provide the facilities.' Abraham Langley, on the other hand, felt they should give some place to the fishing industry views, especially considering the possible advent of a ferry service. 'If we are going to get this ferry,' he said, 'fish will be transported from Stornoway all over the country, to Billingsgate if necessary. We should be ready for the increased fleet of boats, the increased catch and the increased transportation this ferry will bring. It could mean the turn of the tide as far as the fishing industry is concerned, and this board, being responsible for the proper running of the Harbour, should be very anxious to provide these facilities as soon as possible.'

The discussion moved on, covering the type of structure required and the best situation for it. Mr Langley reminded the meeting that the insistence on the pier edge site came from the body to whom the Commissioners were looking for finances. Edwin Aldred insisted that 'even if you put it on the edge of the pier, you can't get every fishing vessel alongside the quay at that point.' Councillor Langley maintained that, 'If you are going to have men working from boats carrying boxes etc to a market 50 yards from the quay, where traffic is running, you are going to be in a fine mess.'

Mrs Urquhart said she was sick and tired of all this talk. 'We have been sitting here for an hour talking about it,' she said, suggesting that the Commissioners should have a meeting with the fishermen. 'I think it is better if people meet rather than having these endless letters from the pundits in St Andrew's House.' Mr Langley suggested they should invite the pundits to the meeting as well. The meeting finally decided, on the motion of Provost Stewart, seconded by Ex-Provost Mackenzie, that the Commission meet with representatives of all the fishermen engaged in white fishing to discuss the provision of a market.

This took place on 15th June 1963 in the Council Chambers. The fishermen were anxious that a fish market should be built to enhance the marketing of fish, and they were unanimous in their belief that such a structure should be at the quay edge, as was the case in other fishing ports. The outcome was that the Commissioners decided to give the pier-edge position some consideration. Captain Mackay was instructed to provide a site plan and came up with a market 50ft long by 20 ft wide situated 8 ft from the edge of the pier. This would encroach on the existing roadway by approximately 4 ft. The Commissioners approved of the plan and agreed to make an application to the Department of Agriculture and Fisheries for grant aid to build it.

The changing fortunes in the fishing industry – the onset of the popularity of shellfish, the shift to larger boats, the shortage of herring – forced larger firms like Duncan Maciver Ltd to take a fresh look at their position. They gave up some of their office tenancies in 1960. Three years later Messrs Coastlines Ltd relinquished their tenancy of Transit Sheds 1 and 2. The Liverpool Shipping Company, which had served the port since 1925, decided they had to stop their weekly cargo service running from Newcastle to Leith, Aberdeen, Belfast and Stornoway. The *Durham Coast* had been their first vessel, followed by the *Welsh Coast*, the *British Coast*, the *Pacific Coast* and others. The cargoes they carried included anything from seed to machinery, all the commodities required for life on the islands. Now, although the Stornoway trade seemed encouraging, the northern run to the other ports had been steadily declining and it was no longer viable to bring a ship around the North of Scotland just for one port. The *Pacific Coast* did the last run at the end of November 1963.

In 1967 the Commissioners bought a mobile crane due to the difficulties in discharging bulk cargoes. The labour force of dockers was dwindling and within another few years would disappear with the advent of the roll-on/roll-off ferries.

Wanted – a fast and efficient Ferry Service

The community as a whole, and two men in particular, put up a considerable fight in the 1960s for a better ferry service with the mainland. As early as February 1960 the *Gazette* was reporting that the Ullapool populace were discussing what preparations needed to be done at their side of the Minch; and also that Ross and Cromarty Council would likely qualify for a subsidy and grant to run a ferry service under the terms of the new Shipping Services Bill going through Parliament.

The idea had been largely promoted in the County Council by Angus Macleod, councillor for the Ranish district of Lewis. He had first suggested that the County Council might run it themselves but this idea had never come to fruition. His next approach was to MacBrayne's, but he reported that they were more interested in promoting a ferry service across the shorter sea route between Harris and Skye. (Despite Councillor Macleod's misgivings about the convenience of this route for Lewis passengers, MacBrayne's went ahead with it.)

The type of vessel to be used on the Stornoway/Ullapool run was a matter of much discussion, some favouring a transport ferry such as those on the English Channel run. Others felt that a tank landing craft, such as the one running from Loch Ryan to St Kilda, would be the answer. The large ferry would be very expensive, and the landing craft would be slow, but either way the vessel should be able to carry around 50 vehicles.

'The complete inadequacy of the vehicle accommodation on the *Loch Seaforth* is generally acknowledged,' said the paper. 'Long waiting lists for conveyance across the Minch form in the summer, and bookings have to be made many weeks and even months in advance.' The Ullapool roads would need to be widened but this was considered not a great problem.

In May 1960 it seemed as if the ferry was not far off. A deputation from the County Council went to see the Secretary of State for Scotland, announcing that 'a private concern' intended to build a new 1400 ton ship, capable of carrying around 50 vehicles and 300 passengers and with a speed of 17 knots. The Scottish Office asked for more details but it was almost two years later before that information was supplied. The Commissioners told the County Council that they supported in principle the idea of a ferry service between Stornoway and Ullapool. The County Council see-sawed between support for the idea, withdrawing in the Spring of 1962, changing their mind again three months later.

In October 1960 they considered a letter from Henry J Gray & Connochie, Advocates, Aberdeen, regarding the inauguration of such a service. They described the intended service as one for merchandise as well as passengers, on a ship as large if not larger than MacBrayne's vessels 'and suitably adapted for all traffic requirements'. It would be similar to the Cross-Channel ferries, they stated, with accommodation for 300 passengers on the upper deck; all the underdeck would be used for motor cars and 'motor vehicles laden with merchandise'. No loose cargo would be carried. They inquired if the Commission would be prepared to consider favourably giving the necessary harbour facilities. Their client, they added, was Mr Charles Alexander, Haulage Contractor, Stornoway.

'Charlie Alec' was a native of Peterhead but had lived in Stornoway for about half a century. He had married Catherine Macdonald from Tong and his children had been born and brought up in the town. Although his family background was in the fishing trade, he had gone into the haulage business and eventually into motor engineering, establishing his own company. He was convinced that the future for island communications lay in a fast ferry service and was determined to see one established.

It was almost three years before any firm proposals appeared, either on the Commission Agenda or in the press. The *Stornoway Gazette* reported in January 1963 that 'a group of businessmen in Stornoway and elsewhere has obtained an option on a Danish car-ferry, the *Isefjord*, with a view to opening a service between Stornoway and Ullapool.' The vessel had 32 cabins and could carry 60 vehicles and 360 passengers. However, the paper reported, there were several difficulties in the way. The Harbour Commission had not yet been officially approached about the ramps which the ferry would need and, although they had always been 'a progressive body, anxious to provide any facilities of benefit to the port,' it was acknowledged that they would be unable to carry out any more major works without Government assistance. That might not be forthcoming, given that the Government, through its contract with MacBrayne's, had interests in both the *Loch Seaforth* and the proposed Harris/Skye ferry service.

The formal approach to the Commission was eventually made in April 1963 in a letter from his solicitors, announcing that Mr Alexander was 'in a position without delay to commence'. Before such a service could begin, however, proper landing and loading facilities for merchandise and passengers would have to be provided at the port. Charles Alexander himself told the Secretary that he intended to use the existing piers to begin with, by serving Stornoway at high tide. A week later he informed his solicitors that he had bought a ferry with accommodation for 50 to 60 cars and approximately 400 passengers.

The Commissioners requested more definite information on what facilities would be required for the vessel to operate the proposed Stornoway/Ullapool run. Meanwhile the press reported that businessmen in the town had been asked if they were interested in investing in a public limited liability company to run the service. The circular letter sent to them had listed the advantages and pointed out that public opinion was behind the idea. 'The cost of transporting merchandise from and to the mainland would be more economic than meantime prevails,' said the letter.

By June 1963, despite several meetings between Mr Alexander, his solicitors, the Commission's chairman and officials, as well as the Consulting Engineers, no ship had been secured. The Board of Trade had not passed the first vessel as suitable and the company was negotiating to buy a German vessel but could not conclude the purchase until they were sure it would be passed by the Board of Trade. If this proved not to be the case, then offers were being taken for a new vessel which, if proceeded with, should be ready by the summer of the following year. Even if he had to build a new ship, he still intended to go ahead with the service in the meantime, using a chartered vessel. This should be in place before the end of the year. Until the necessary pier facilities were completed, he intended to operate at both Ullapool and Stornoway 'at stages of the tide to permit the use by the ferry of the existing quays.'

A prospectus from the new company, Ullapool/Stornoway Ferries Ltd, was received in July 1964 but in May 1965, Charles Alexander died suddenly. For almost three years he had been struggling to realise his dream of an Ullapool to Stornoway ferry connection. With his death Ullapool/Stornoway Ferries Ltd was dissolved and the focus returned to MacBrayne's.

Macbrayne's weekly cargo vessel *Loch Dunvegan* arrived from Glasgow every Friday in the 1950s and 1960s

Mrs C Maciver

HM Queen Elizabeth shown herring packing accompanied by Harbour Commission Chairman Norman Maciver, 18th August 1956

Michael Maclennan

Stornoway pilot boat *Portrona* leaving the Drill Ship *West Navion* and her accompanying tugs 14th December 2001

The Age of Oil and Ro/Ros 1970s–1980s 8

THE FIGHT for the ferry, however, was almost won. The Scottish Transport Group did come down in favour of the provision of the ferry link and stressed that it was their firm view that 'there could be no point in providing this service other than as a complete modern facility offering roll-on/roll-off transit for maximum weights.' They arranged a meeting in Dingwall of all interested parties for 3rd February 1970

The Chairman, Bailie Alasdair M Matheson, along with fellow Commissioner Donald Macleod and the Harbourmaster were delegated to attend for the Commission. They reported back to the Commissioners a week later that the new ferry could be on the cards by the following year. However, if the cost of development at either side of the Minch exceeded £100,000, the scheme could be delayed by a further twelve months.

'It would be very wise,' said Donald Macleod, 'to give this our full support with the hope that it will get going in 1971.' The decision was taken to erect the necessary pier installations, subject to grant and loan aid and the Secretary of State's approval. The Commissioners also approved of the Transport Group's offer to undertake the initial investigation into the pier requirements at both ports.

Two weeks later, Malcolm Morrison from the Consulting Engineers Archibald Henderson and Partners gave a special meeting the report of his first survey. MacBrayne's intended to inaugurate the service with the MV *Iona*, which they were in the process of building. This boat would be 230 ft with a 44 ft beam, but the company had asked Mr Morrison to make provision for a larger vessel, 250 ft with a 50 ft. beam. He advised the construction of a 72 ft ramp near the 'bend' of King Edward Wharf and also stressed the need for more parking facilities near the ferry berth. At their March meeting the Commissioners considered his proposals on how to arrange the parking. The demolition of Transit

The Old Fish Mart

A stumbling block in the way of all the progress of the new ferry terminal was reported to the April 1970 meeting of the Commission. Mr George Pease, the County Council Planning Officer, while approving the application for the construction of the ferry terminal, made it a condition that the Old Fish Mart were retained. The Commissioners replied that 'the building was no longer of any architectural merit and was structurally unsafe.' They also lodged an objection with the Secretary of State for Scotland, saying that they had a duty to make use of the pier facilities to the advantage of users of the harbour and the community .

The demolition of the old Mart did not go down well in many sections of the community. A year earlier the *Gazette*, had urged a 'Save the Mart' campaign, quoting the example of Charter Oak School in the American mid-West, an eight-sided brick construction built in 1872 and now preserved as a historical monument.

'If the Stornoway Fish Mart, an eight-sided building unique to Lewis, if not to Scotland, must not

The new ferry terminal, opened in 1997, was designed to reflect the octagonal shape of the old Fish Mart ; (Opposite) from a painting by Laura Macfarlane

stand in the way of progress, is it physically beyond the ken of machines and men to have it moved to a spot where it can become a focal point for the visitor's cameras as, say, an attractive museum? . . .'

The Lewis Amenity Trust had taken up the cause, suggesting it be retained as a home for the Lewis and Harris Folk Museum. They felt that the building 'was a landmark which was regarded with considerable affection and should, if at all possible, be retained because of its association with Lewis history.'

Not everyone had the same feeling for the old place. Mrs Urquhart, addressing the County Council's Planning Committee in July 1971, told them that 'the Pier and Harbour and the Town Council, the Police and MacBrayne's, all want this 'thing' removed – which resembles for all the world an ancient prison with a central exercise yard and cells branching off.'

The Committee capitulated and the demolition was sanctioned.❑

Sheds 1 and 2 were agreed unanimously but Mr Morrison said this did not leave enough room for off-loaded vehicles. To provide sufficient space either the old Fish Mart or the old Harbour Office at No 3 South Beach would have to be demolished. This decision went to a vote and the old Fish Mart lost out.

The initial production of plans and costings for the new terminal did not go smoothly. Due to confusion between the Consulting Engineers and the Scottish Transport Group over the specifications required for the installation, it took three estimates over the summer of 1970 before the final costing figure of £95,759 was arrived at. Henderson's had always been concerned that these specifications appeared to be geared towards accommodating the *Iona* which they said had an 'unconventional bow ramp'.

They also felt that installing fendering which would allow both bow and stern loading was an over-provision. After direct consultation with Mr W M Little, manager of the Scottish Transport Group, they reduced their plans considerably but appeared to remain sceptical of the practicality of the Group's ideas.

Mr Little, stung by press reports of the Commission's concern over the changing costs, said he was sure that the Commissioners would not wish a terminal designed for a single ship and this was not his Group's intention either. Indeed, they were also in touch with Western Ferries who owned the only three roll-through ferries operating on the West Coast. 'We all should be able in emergency to use any terminal,' he added, urging the Commissioners to apply to the Scottish Development Department for grant aid.

Finally, in January 1971, the Secretary of State Gordon Campbell announced that he has approved the inauguration of a ferry service between the two ports, which would be eligible for 75% grant assistance. The Scottish Development Department were prepared to 'consider sympathetically' a 25% loan, depending on the Commission's financial position. They suggested a meeting in Stornoway to finalise all the details.

While all the technical and political disputes had been going on, opposition to the Ullapool link was steadily growing on the island, particularly in the business community. Rumblings had begun in the ranks of the District Council in August 1970, with some councillors worried about the loss of the Kyle connection and the withdrawal of the direct cargo boat from Glasgow. At the end of 1970, several local business men formed an action group to fight the proposal.

•

The *Iona* came under fire on her first crossing to Kyle. She rolled so much in fair weather that it was doubted she would cope with a gale. Livestock had to be left on the pier at Kyle and those animals aboard were so tightly packed that two did not survive the crossing. After a week, the Council asked for talks with the Secretary of State with a view to having her replaced. After two weeks the Chamber of Commerce was referring to the service as 'chaos'; by July the Provost was urging people to make their complaints directly to MacBrayne's; and in November the Council took their complaint to No 10 Downing Street.

In December it was announced in the press that the car ferry *Clansman* was in dry dock being lengthened and fitted with roll-on/roll-off facilities. The Secretary, writing to the Scottish Development Department, expressed the Commission's annoyance that neither they nor the Town Council had received official written notification of this plan.

By this time David MacBrayne Ltd had ceased to be the operator of the West Coast ferries. From 1st January 1973, the Commissioners were dealing with the new company of Caledonian MacBrayne Ltd, an amalgamation of the shipping operations, both of the old company and those of the Caledonian Steam Packet Company. Their General Manager, Mr N J D Whittle informed the Commissioners that the link span and the extension to the pier at Ullapool were now almost complete and they expected to be able redirect the present ferry there in February.

'While this will enable the benefits of end loading facilities to be derived at the Ullapool end of the route, the full benefits cannot of course be realised while it is necessary to continue with hoist operations at Stornoway,' he added. The *Clansman* would be ready in April but, having no hoist, could not be deployed if Stornoway's link span was not operational.

The link span was not, in fact, ready for the April deadline; bad weather and a trawler collision with the new dolphin put paid to that. Graydon Construction did not start to install the link span until 9th April 1973.

The link span was finally tested and passed for use at the end of May 1973 and Graydon Construction arranged for one of the harbour staff to be instructed in its operation and maintenance. The *Clansman*, however, had missed her deadline and was still in dry dock, so it was the *Iona* , watched by a large crowd, which inaugurated the service. Provost Matheson was on the quay to welcome the first car to drive off, that of Captain Angus Murdo Macleod and his wife Frances, returning to set up home in the island. Captain Macleod, better known in the town as 'Texie', would a few years later become Deputy Harbourmaster at the port, and later still, first Harbourmaster of the new Stornoway Port Authority.

Despite all the celebration – and relief – the new ferry service seemed destined to be troubled. On 13th June, a fortnight into the service, one of the supporting pawls on the link span did not engage and the ramp fell, shearing a bolt but fortunately causing no injury. This worried the Commissioners enough to begin looking into the possibility of a reserve ramp, in case of future breakdowns. Caledonian MacBrayne were concerned, since if there were a similar occurrence when the *Clansman* took over, there was no other way of discharging the ship.

The *Iona* ran for the best part of a week on one engine, then succumbed to a broken crankshaft. She was relieved by the *Columba*, but this vessel being required for another area, there was no ferry between Stornoway and the mainland at the end of that week. A joint statement from John Macleod, Chairman of the Harbour Commission and Provost Matheson said they were horrified to learn of the difficulties experienced by the travelling public at Ullapool. They sent a telegram to MacBrayne's asking that a hotel lounge and other facilities be provided at Ullapool for those awaiting the *Clansman's* arrival from Stornoway on the Saturday morning. 'We insist,' said the telegram, 'repeat insist – that it is your clear duty and responsibility to make immediate provision for shelter, feeding and toilet facilities at no extra expense to those travelling.'

Meanwhile, the *Clansman* , making her way north to the rescue, sailed into Stornoway harbour on Saturday morning and hit the *Iona*, lying at the quay for repairs to the offending crankshaft. It was four o'clock that afternoon when, after minor repairs, she sailed for the beleagured and frustrated travellers at Ullapool. Both the *Clansman* and the Harris ferry *Hebrides* did emergency crossings on Sunday to clear the backlog of vehicles.

The Town Council, up in arms over the 'catastrophic situation', decided not to accept the invitation of the Scottish Transport Group to the inauguration ceremony being held aboard the *Clansman* on 6th July, which was to be performed by the Scottish Secretary, George Younger. It would not be proper, they said. The Harbour Commissioners played safe, perhaps aware of their dependence on the goodwill of the Scottish Office, and 'in view of the fact that the Commission were not directly involved and did not have he same responsibility in this matter as the local authorities,' left it to the individual Commissioners to decide whether or not to attend.

That winter of 1973-74 the protests succeeded in getting a promise of a new vessel to replace her and in the Spring of 1974 the *Suilven* was launched in Norway. This new vessel had double the carrying capacity of the *Clansman* and was a much heavier vessel than her predecessor. Her first run, in August 1974, should have been the official inauguration with only invited guests on board but a breakdown of the *Clansman* at Ullapool meant it was a proper operational crossing, loaded with passengers and cars. The celebratory run took place the following day.

Re-organisation of Local Government

There were three events in the 1970s which drastically changed, not only life at the harbour, but life for the whole of the island community. The first of these was the advent of the roll-on/roll-off ferries, quickening its pace and making it more accessible to many. The other two elements were the growth of the oil industry and the re-organisation of local government.

In the autumn of 1969, the Government's Wheatley report on the Re-organisation of Local Government was published and at the end of 1971 the Secretary of State Gordon Campbell announced that the Western Isles from Lewis to Barra were to have the same special status as the Orkney and Shetland Isles. The idea met with approval in Stornoway where the local bodies had been advocating such a move for years. The reorganisation dissolved the ties of Lewis with Ross-shire and those of Harris and the Southern Isles with Inverness. The whole chain from the Butt of Lewis to Barra Head was to be administered from Stornoway by a council made up of representatives from every district in between. Stornoway Town Council would cease to exist after 15th May 1975.

Stornoway Pier and Harbour Commission, on the other hand, was an independent body, established by Act of Parliament, and therefore not under the control of any council. It was not, however, unaffected by the changes taking place in the other spheres of local government, since by statute three of its members had to be Town Councillors representing wards within the old Burgh of Stornoway. With the disappearance of Stornoway Town Council and the emergence of new wards, the rules had to be looked at again. In March 1974 Murdoch Macleod (who had succeeded Alasdair Macleod in 1968) reported on the need for the Constitution to be amended to allow the new Western Isles Islands Council to appoint three of their number for a three year term.

Since they were already looking for a Provisional Order to extend the harbour limits, the Commissioners took the opportunity to deal with this, as well as making themselves the Pilotage Authority for the Harbour; adding a prohibition of anyone carrying out any works in the harbour without a licence from themselves; and the powers to carry out the extension of No 2 Pier.

Apart from the constitutional changes, the re-organisation affected the administration of the Harbour Commission and at the end of 1973, with the change-over to the new Western Isles Council drawing nearer, the Commissioners agreed that they needed to have their own full-time official in place. They decided to employ a General Manager, who would take on overall control of the harbour administration, as was the case in the new harbour authorities of Cromarty Firth and Shetland Islands and already existed at Lerwick Harbour Trust. The duties would include those of Secretary and Treasurer, as well the control of all the Harbour staff and the guidance of the Commission on development, finance, and all aspects of the harbour operation.

The Commission offered the post to their part-time Secretary/ Treasurer, Murdoch Macleod, who took up his full time duties on 16th May 1975. The post kept the title of Secretary/Treasurer until the end of 1977 when the designation of General Manager was added to it, bringing into line with the procedure in similar ports.

The Commission had always rented part of the Town Clerk's office in the Town Hall for his Harbour Secretarial duties and also used the Convening Room there for their monthly meetings. With the change in administration these premises were no longer available. The monthly meetings held the Harbour Office at 4 South Beach over the next few years involved much shifting of furniture beforehand. Desks were pushed together to make a large 'table' and Commissioners of various girths squeezed themselves into chairs around it elbow to elbow. Moving a chair back meant bumping into the long counter that ran almost the length of the room. It was not ideal and it gave a push to the drive for new offices which had been on the back burner.

At the beginning of 1975 the Commission took the decision to convert Amity House into offices and welfare facilities for the dock workers but over the next two years these plans took a different direction. In August 1977, after a meeting with the Highlands and Islands Development Board, it was agreed to make Amity House the new Harbour Offices; while the dockers' facilities, public toilets and watchroom would be accommodated in the existing watchroom block at 3 South Beach, suitably renovated. The go-ahead was given the following May when funding for the total cost was settled as grant from the Highlands and Islands Development Board and loan from the Royal Bank of Scotland.

The work was carried out by John G Macgregor (Stornoway) Ltd, whose carpenter Peter Dyer crafted the long table for the Meeting Room, the smaller version for the Chairman's Room, and two side tables. He also made the set of lockable wall cabinets in the Meeting Room for the individual members to keep their papers, and a cocktail cabinet for the Chairman's Room, shaped in the manner of an old sea chest. The Chairman's gavel was carved in the shape of a miniature capstan and its block into a coil of rope. The chairs in all of the first floor rooms were covered in Harris Tweed, and these have remained almost as good as new over the intervening thirty years. On the suggestion of ex-Provost Matheson, it was later agreed to approach Comhairle nan Eilean to obtain an indefinite loan of pictures relating to the harbour, and thirteen items were received in the summer of 1979. The Harbourmaster had weather instruments fitted in the hall of Amity House so that the harbour users could check wind speed and direction and temperature.

The work on Amity House was completed by the end of 1978 and the opening ceremony was performed on 6th December by the Deputy Chairman of the Highlands and Island Development Board, Rear Admiral Dunbar Naismith. He paid tribute to those who had served on the body down through the years and without whom 'the present day facilities which are evident in this magnificent harbour would not have materialised.' He noted that members of the Commission had never been a 'push-over' for anybody in the past and that it seemed the present-day members were following in their footsteps.

In his speech Rear Admiral Naismith was enthusiastic about the future of the Western Isles fishing industry, expressing the belief of the Development Board at the time that this lay in the under-exploited grounds west of the Hebrides and in boats 'much bigger than any of us could visualise not so long ago.' However, he also alluded to the development of wave power, saying that the area off the Western Isles was considered to be the most promising in the UK for this, but would require enormous investment and was at least thirty years away. Declaring Amity House open, he complimented the Commission on having the vision not to bulldoze the building and erect a concrete and glass box in its place. Posterity would be grateful, he said, and he ventured to suggest that Amity House would be standing long after the glass boxes had disappeared. Referring to the fact that the building took its name from the old barque which had been moored across from the front door, he said, 'Accordingly, I hope that it is appropriate to ask God's blessing on this House and all who serve in it and with these words I have great pleasure in formally declaring your new office premises well and truly launched!'

The Oil comes

The fast-expanding oil industry was a major factor for change in the Highlands and Islands, as well as in Stornoway. At the end of 1972 the Regional Development Division in Edinburgh wrote to the Assistant County Clerk detailing the port requirements for the servicing of oil rigs. The letter, forwarded for information to the Commissioners' final meeting of 1972, said the vessels would need berthage between 160 and 250 ft long, and would require a depth of 21 feet of water. The quays would have to be able to take heavy loads and there was usually a need for warehousing close by. Each rig would have two or three supply vessels servicing it, which would operate at any time of the day or night, including Sundays. An estimate of the supplies used by a rig over a year was 25,000 tons, with 50% being water and fuel, 25% mud and cement, 15% tubular steel, and 10% miscellaneous, including catering.

Interest in the port was hotting up a bit as oil companies began to explore the seas west of the islands. Enquiries started to come across Murdoch Macleod's desk with all kinds of queries and offers, from the exploration company who offered a loan towards pier construction to the gentleman in London who wanted to reclaim twenty acres of Newton Basin, on a 200 year lease, for a nominal rental sum but was 'happy to discuss any other terms which might be attractive to the Commission.'

The Commissioners began to consider the management of their land in the light of the developing industry. One of their first actions was to look into the cost of extending the existing piers. A report by Archibald Henderson & Partners, suggested that No 1 Pier could be made into a structure capable of handling the bulky traffic of the oil industry. They estimated it would cost £365,000 to extend it by 300 ft.

Meanwhile Stornoway Trust Estate had been approached by the Norwegian firm of Fred Olsen Ltd, who were keen to establish a £6m development at the port. On 26th September 1973, they held a special meeting, to which representatives of the Commission were invited along with some from the County Council, the Department of Employment, the Highlands and Islands Development Board, the Northern Lighthouse Board, the Ports Directorate of the Department of the Environment, and the Regional Development Division of the Scottish

Office. Also present was the Managing Director of Fred Olsen Ltd, Michael Thompson and other representatives of the firm.

Fred Olsen Ltd, like its parent company in Norway, was involved in the construction of oil rigs and other craft, and they were interested in constructing a service base at Arnish Point. Mr Thompson said the company would 'fit in with the community' by employing local labour, hopefully attracting people back home to the island. They envisaged a maximum workforce of 1,000 through training and apprenticeship schemes. The development would be in three stages – first a supply base and a construction unit for simple types of structure; then a unit for the more sophisticated type of structure such as production platforms with concrete bases and steel decks; and finally a dock in which to build sections of vessels and pontoons for semi-submersible rigs.

The Trust had to decide whether or not to give the land to Fred Olsen Ltd for a development which would change the face of that part of the Harbour, involving the blasting away of a small hill. Murdoch Macleod was invited to accompany some of the Trustees to Norway to view Fred Olsen's operations in Norway, visiting an oil rig supply base at Tananger, close to Stavanger, a shipyard on the island of Stord, and the company's larger shipyard at Bergen, all of which gave a favourable impression. The Secretary, along with Commissioner Kenneth Maclennan, had also accompanied the Trustees to Shetland where Fred Olsen Ltd leased thirty three acres from Lerwick Harbour Trust, in order to see how the authorities there had handled the development.

There were statutory implications for the Commissioners in this development. For one thing, although the land and foreshore at Arnish already belonged to Stornoway Trust, the seabed on which Olsen's would construct their pier belonged to the Crown Estate Commissioners. For another, Stornoway was still classified as a 'marine work' under the terms of the Harbours Act 1964, and therefore the Scottish Office had a say in its constitution and legislation. The Ports Directorate advised that the Commissioners consider extending the limits of the harbour to take in the Arnish peninsula. The Commission's lawyer, Colin Scott Mackenzie, engaged to handle the negotiations with the bodies involved, confirmed that the Crown Estate were willing to sell the seabed to the Commissioners.

The question was whether they could afford to buy it. The Trust, on the other hand could. Fred Olsen Ltd were in a hurry to start by the projected date of 16th May 1974 and were reluctant to have to negotiate agreements with more than one authority, or, indeed, to have two landlords. The main problem facing the Commissioners was that they were already committed to several large projects around the harbour, which required funding. They had no idea what purchase price the Crown Estate would set on the seabed and they were afraid of overstretching their borrowing powers by committing to it.

On a 7 to 2 vote the Commissioners agreed to the proposition that they would make over the right of purchase to Stornoway Trust. This would mean there would be only one agreement to be negotiated, that between Fred Olsen's and the Trust. That agreement would contain assurances for the Commission on two main points, namely that harbour dues would be paid to them for ships berthing at the pier there, and that the pier would revert to the Commission if and when the company left. A sum of one tenth of the purchase price was paid by the Trust to the Commission for having conveyed the purchase rights to them.

A buzz was beginning in the community with some local businesses expanding to meet the needs of the new industry, others finding it difficult to get staff as their employees moved to new jobs. The Council began to think of providing new houses and schools for the expected increase in the population. A joint delegation of Trustees and Councillors visited Shetland in May to view at first hand how the industry was affecting these islands and how the problems had been met. The Commissioners decided in August 1974 to take up the Babtie proposal for the extension of No 2 Pier by 450 ft at an estimated cost of £770,000. Although they were able to carry out a site investigation, no funding was obtainable for this project for years.

At the end of 1974 the new company of Lewis Offshore Ltd was awarded a £4m grant from the EEC to 'help establish in the Isle of Lewis. . . a steel fabrication complex for offshore structures.' In October 1974 the new base came a step closer when Explosives & Chemical Products Ltd asked for permission to moor the *Isle of Skye* in Glumaig Bay as 'a floating magazine' carrying the explosives needed to level off the land at Arnish. Blasting operations at the site began in early 1975 and the earth was used as reclamation for part of the harbour at Glumaig. The local fishermen lodged an objection, worried about danger to the fleet entering and leaving the harbour. At a special meeting attended by HM Inspector of Explosives, they were assured that it was the handling of explosives which had to stop not the traffic.

In the Spring of 1977 the Secretary and Commissioner William John Macdonald, along with the Factor and Chairman of Stornoway Trust, visited Aberdeen to look at recent oil developments there. They reported that, while there was likely to be some activity in the waters off the west coast within the next five years, it was unlikely that Stornoway would be anything more than a supply base. There was an opportunity for firms to benefit from the supply business if they wanted to take the opportunity.

Between them the Commission and the Trust formed a Working

Party, composed of three members of each body, along with their respective officers, to do what they could to encourage the oil industry to look at Stornoway and the island. The Working Party were to draw up a policy for harbour and land utilisation for potential oil business in the area; to promote these interests for the benefit of the community; and to visit any areas or organisations which might assist them.

The Drillmaster

Drillmaster

The biggest of all Lewis Offshore's jobs took place in 1978-79 when the Drillmaster rig was towed to Stornoway to be converted from a drilling rig to a production platform for BP's Buchan field. Work at the yard had dried up, and it had been put on a care and maintenance basis in the summer of 1978, in spite of the best efforts of the Council, the workforce and the MP to avert this. The contract entailed some dredging at the entrance at Glumaig for which permission was requested in May 1978, but although the result of the Company's negotiations and the award of the contract were common knowledge in the community and even announced in the press, no official approach was made to the Harbour Commission about arrangements or charges until shortly before the rig was due to arrive in October.

The official application came before their meeting on 9th October but they deferred consideration until more information was available to them. Lewis Offshore informed them that, as well as the rig, there would be the *Lonka* (in use as a floating site office), two flat top barges, a hatch barge, a floating crane, and daily ferrying movements of men and material from the quay to the rig. The Commissioners were angry that, although the company had known for six months what was entailed, it had been left until the last minute to apply for the use of Glumaig Bay. As they met, the *Drillmaster* sat at anchor off the Briaghe waiting for their decision, and there was a feeling round the Commission table that they were being railroaded into decisions that would have extremely important implications for the harbour's present economic situation and for future negotiations.

The Commissioners, therefore, agreed that the use of Glumaig Bay for this operation must be charged at a full commercial rate and set the dues for all vessels involved in the application at £25,000 for the period to 31st May 1979 or the end of the project, whichever was the earliest.

The charge was non-negotiable and they drew up an Agreement with seventeen conditions, which had to be accepted before any work could begin. They did consider not allowing the rig to enter the harbour but decided against that, insisting only on the work ban.

Commissioner Sandy Matheson told the *Gazette* why the Commission had taken this line on the matter, explaining that they had to finance capital expenditure of more than £300,000 for repairs to piers, a new fish mart and other projects. This had to be found from the duespayers such as the shipping companies, fishermen, ferry passengers and other harbour users. Pointing out that the *Suilven* at less than 2000 tons paid £10,000 dues per annum, he added:

> 'If we let the Aker Group off with nonsensically low rates the burden will have to be carried by local duespayers which is unfair and bad commercial practice. I see no reason why local people should subsidise through harbour dues the profits of this or any other company.'

The company (who had priced for harbour dues at £3000) replied that they intended to appeal to the National Ports Council against the level of dues charged, although they offered to put the whole £25,000 in the Commission's bank account pending the result of this appeal, in order that the beginning of the work should not be delayed. The Secretary contacted Michael Thomson, Lewis Offshore's managing director, and asked for an assurance that no work would take place at Glumaig until they accepted the terms of the Commission's licence.

A further application came before them in December 1978 from Messrs Aker Offshore Contracting Ltd, who requested a berth for the passenger vessel *Borea*, which was to be used as floating accommodation for around 300 men. The Harbourmaster pointed out that this ship could only use the deep water berth on the east side of No 1 Pier, which in the previous year had been used by about 60 vessels. Three deep water anchorages in Glumaig Bay were no longer available to him because of this contract, and therefore Captain Mackenzie could not afford to lose this berth for the extended period of twenty weeks.

The vessel was due to arrive in the port on 24th January 1979. The Commissioners held a special meeting on the 22nd to consider Aker Offshore's intentions. They agreed that they could not deal with any application until it was clear that Aker were making a formal application or were simply telling the Commission what they intended to do. The company replied that the men were additional workers taken in to complete a contract by a set date; extensive searches had been made to find enough accommodation, with no success; and they had not realised they must make a formal application for mooring the vessel. Having rectified this, the company accepted the terms of a formal agreement drawn up Bird Semple and Crawford Herron.

The Commission set the dues for this operation at £20,000 and issued a press statement, saying that permission was given on this occasion only because of the difficulties in which Aker Offshore Ltd found themselves. It was unlikely that permission would be given again because it was not in line with the Commission's thinking on how the harbour should be developed. They were warning the company that no future applications would be considered unless they were submitted 'timeously and properly' and a meeting was being arranged with representatives of Fred Olsen Ltd to discuss matters. An assurance would be required that 'there will be no repetition of their regrettable lack of co-operation with a statutory authority.'

The meeting took place on 6th February 1979 with Michael Thompson and George N Macleod representing Lewis Offshore Ltd. Murdoch Macleod put six questions to Mr Thompson regarding why, before the Commission had been informed about the tender for the *Drillmaster* conversion, the contract had been let with the company using the Commission's facilities.

Mr Thompson replied that Lewis Offshore was under an obligation to their client to observe secrecy. The Commission would have to realise, he said, that they were dealing with construction people who suffered great pressures and many problems. Every contract in which the company would be engaged in the future would require use of the Commission's facilities and the Commission would have details of these as soon as it was commercially reasonable to do so.

This brisk approach was dealt with swiftly by the Commissioners, starting with ex-Provost Matheson who expressed his dissatisfaction with the Company's behaviour throughout, and his feeling that they had failed in their relationship with the community. He suspected that Lewis Offshore had attempted to force their plans through, which meant that from now on everything the company did would be scrutinised with the greatest care. He was equally dissatisfied by the discussion at that particular meeting.

Michael Thompson backpedalled and agreed that the Commission had not been properly consulted; the company could have improved communications and he would attempt to correct that. However he felt that the Commission were also being very inflexible in their attitude.

Commissioner William John Macdonald closed the discussion by expressing regret that such a large organisation with whom the Commission were dealing, should be submitting such minimal information. The Commission, he said, were used to dealing with people who gave clear cut statements of their requirements and then no one could say that the Commission were unjust or unfair in their dealings.

When the contract had to be extended into the summer of 1979, Michael Thompson wrote to thank the Commissioners for the co-operation and courtesy shown, both to him personally and to his company, in approving the extension. The *Borea* was replaced by the much larger *Najla* at the end of May 1979, to take the 450 workers who were required to finish the work on time. The Firemaster kept a close watch on safety arrangements, as did Offshore Supplies Office of the Scottish Development Department in Glasgow, passing their comments on to the Commission via the Harbourmaster. At the height of the contract in July 1979, about 1300 people worked at the yard. The contract lasted much longer than expected, delayed by materials shortages and labour troubles, and the *Drillmaster* became such a familiar sight in the harbour that she was missed when she finally left on 13th September 1980.

Angus Smith

The Arnish development showing the Drillmaster and the accommodation vessel *Najla* in Glumaig Bay

The heavy lift vessel CEC *Crusader* loading steel piles at Arnish pier 22nd September 2003

Michael Maclennan

Stornoway Harbour pilot boat *Portrona* alongside the French fish factory vessel *Joseph Roty*, 6th May 2003

MFV *Succeed* SY186, entering the Inner Harbour. Formerly *Providence* SY345, this vessel was one of 12 boats built under the Outer Isles Fishery Scheme from 1961 to 1963 which aimed to revitalise the industry, as the herring declined. She was decommissioned in 2003

Michael Maclennan

Prawn boxes in the Fish Mart – Landing the catch at Stornoway Fish Market, June1999. Prawn and other shellfish are now the main species fished in the Minch

Stornoway is a regular port of call in the summer months for cruise liners of different nationalities. Here *Song of Flower*, a regular visitor, berths at No 3 Pier

Freight transport systems have evolved over the years, from coastal shipping to short sea crossing roll-on/roll-off ferries – from slings and derricks to hydraulic linkspans

2 photos Michael Maclennan

Charles Macleod

Stornoway has a large fleet of local yachts, as well as many pleasure craft from different countries visiting during the summer months

An annual event of the past decade has been the Sail Hebrides Maritime Festival, whose principal aim has been to introduce young people to the joys of sailing

2 photos Michael Maclennan

Some of the smaller vessels in the Stornoway fishing fleet alongside Cromwell Street Quay

2 photos Charles Macleod

Helen Mackenzie

Now a car park, South Beach Quay once thronged with hundreds of women working at the herring. This bronze figure by Charles Engebretsen and Ginny Hutchinson commemorates them, while in the background the ferry terminal, built in 1997, mirrors the shape of the 1894 Fish Market

J J Maclennan

Naval ships have often been a presence in the harbour – in its days as a military base during two World Wars and as a rest-and-recreation port of call in peace time. This photograph shows the Port Authority's Chief Executive, John J Maclennan, in an off-duty moment inspecting some of the Norwegian fleet in the summer of 2006

Angus Smith

Stornoway ferry terminal from the air 14th May 1997

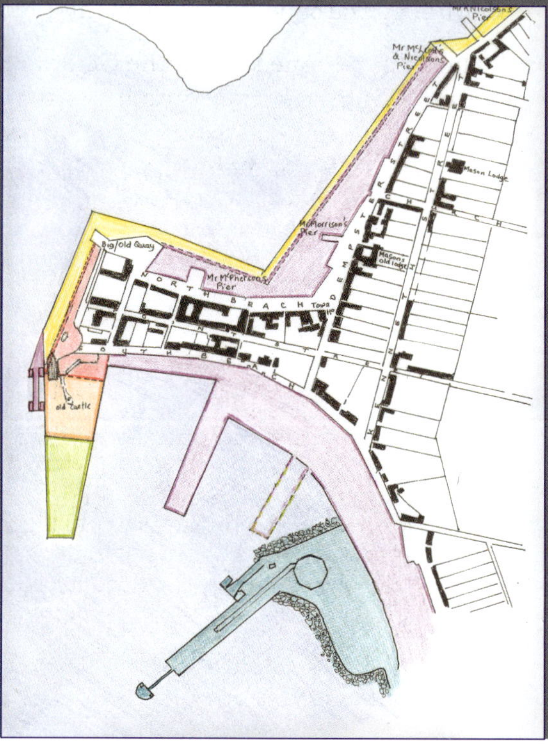

The coloured areas on this plan show the different stages of development of Stornoway Harbour from 1865 until the present day
Red – 1865
Orange – 1882
Lilac – 1892
Yellow – 1930s
Green – 1947
Purple – 1972
Blue – 1997
Not shown – causeway and slipway 1950s

A Take-Over Bid 1982–2004 9

THE COMMISSION were confronted with the possibility of absorption into the local authority when, in 1982, Comhairle nan Eilean put forward their submission to the Committee of Inquiry into the Functions and Powers of the Islands Councils of Scotland, chaired by Sir David Montgomery. The submission had been drawn up by a sub-Committee of the Council, and was looking for extensive powers, including the takeover of the Health Board, the airports and the ferry terminals, the police, fire service and the valuation services. Not all the Councillors were in full agreement with the proposals and the takeover of the police and fire services was rejected by the full Council meeting. However, the operation of Stornoway Harbour remained on this radical wish-list.

When this became known, the General Manager was instructed to write to the Montgomery Committee asking that the Harbour Commission be allowed to present evidence. Stornoway Fishermen's Co-operative was first off their marks to protest directly to the Council, expressing their grave concern at the proposal. They added that:

> '. . . the rapport that existed between the Stornoway based fishing fleet and the Harbour Commission was the envy of fishermen in the rest of Scotland.'

They doubted that this very happy relationship would be improved by a takeover by any other authority. The local press also came out against the plans amid talk in the community of a 'Big Brother' council.

With the glowing recommendation from the fishermen behind them, the Commission went on to appoint a Sub-Committee to prepare the submission to the Committee of Inquiry. On 29th March 1983 Sir David and his Committee visited the port. They were shown round the harbour and put questions to the Sub-Committee, comprising Commissioners Malcolm Gillies and John M Morrison, General Manager Murdoch Macleod, and their legal advisers Malcolm Maciver and David Miller from Bird Semple and Crawford Herron. Malcolm Gillies, reporting on the meeting, said it had been a most useful discussion.

The findings of the Montgomery Inquiry came before the Commission in June 1984 and came down in favour of the Commission. Sir David stated that, having visited Stornoway Harbour and spoken to some of the Commissioners, they believed that:

> 'The harbour is considerably more than a ferry terminal. It is a general purpose port with a large slipway for maintenance and repair. There is evidence, for example from Shetland, that it is possible for a large independent authority and a local authority with harbour powers to operate side by side within the same area. Although Stornoway does not enjoy the prosperity that Lerwick is currently experiencing, we do not think there is any need to abolish the Commission. They are a body whose membership is drawn from local people, representing both harbour users and others with an interest in the operation of the harbour. We believe the harbour is at present being managed satisfactorily under the independent harbour authority, and we do not think that any significant benefit would arise from a transfer to the Council. . .'

The Montgomery Committee did, however, recommend that the Commission consider widening the area from which they drew their representatives of the community 'so that these can be representative of the whole area which relies on the services of the harbour – not just the area of the former Burgh of Stornoway.'

Protecting the fishing

During the first few years of the 1980s the Commissioners were concerned by the troubles of the rapidly changing fishing industry. The entry into the Common Market, the failure of the member states to achieve agreement on a Common Fisheries policy, the herring ban, and the increase in 'industrial' fishing all had an impact on the scene round Stornoway harbour. There were more and more instances of vessels being charged with illegal fishing, arrested for working within the fishing limits or for the wrong type of fish or using the wrong size of net. In the summer of 1977 alone four Spanish trawlers were fined a total of almost £70,000. The increased number of fishing boats registered in the UK but carrying a Continental crew, was a particular annoyance to British fishermen

In the 1980s the trade was becoming heavily industrialised. Fishing vessels were large, with enormous catching capacity, and often landed to factory ships which processed the fish aboard and carried it home to France, Germany or Russia for sale. One such ship was the *British*

Viscount which asked for the use of Stornoway harbour in 1980 as 'a factory ship in a stationary position'. Owned by British Oceanics Ltd, and on charter to Messrs Jiltro Ltd, the intention was that the vessel would take fish from the local fleet using the stern loading facilities. The fish would be processed aboard, then shipped out by ferry. The company hoped to employ only local labour and to purchase all supplies and services locally.

The majority of the local fleet were in favour of the venture, on the grounds that prices would be steady, but the fish processors were worried about their supplies, if the new company did not buy in the open market. The Commissioners held two special meetings to debate this proposal and, on a five-four vote, decided to refuse it. Having committed themselves to the expense of a new Fish Mart, they were afraid that the money would be wasted if the factory ship operation became a monopoly, taking all the fish except for prawns. They stressed that they were not protecting the other buyers in the port, and would in fact welcome more competition at the market if it came from a land based source.

The fishermen did not take this well, threatening to withhold their harbour dues, suggesting to Jiltro that they should moor the ship in a sea loch close to the port, and demanding a meeting with the Commission. This was rejected by the General Manager, Murdoch Macleod, who said that the Commission could not meet under the threat to withhold dues, since they could not negotiate with any party where there was a possibility of legal action pending.

The Commissioners decided, as a result of this episode, that they would set up a Fisheries Committee from May 1980 to try to assist the industry in whatever way possible and particularly to attract fish buyers to the port. They also began to liaise with Comhairle nan Eilean's Joint Consultative Committee on Fishing to discuss the Council's Fishing Plan.

The lack of control over the mackerel fishing was taken up by Mallaig Harbour Authority over the late 1970s and early 1980s as concern grew about depleting stocks and loss of revenue to fishing harbours. Mallaig had continually urged the Government departments to have only three ports on the West Coast designated for this fishery: Mallaig, Stornoway, and Ullapool and that all catches should be sold by auction at these ports. Mr Thom, Mallaig Harbour's Secretary, protested to the Secretary of State that the harbour authorites of the three ports were concerned that Government's unwillingness to exercise effective control would lead to the early demise of the mackerel fishery. The Commission supported Mallaig's frequent protests.

The Government's failure to get agreement on west coast herring with the EEC partners appalled the fishing industry in the autumn of

1981. Harbour Commissioner James Macrae observed that some of the delegates to Brussels seemed to him like kippers. 'They are without guts', he said. Mallaig sent a telegram to Margaret Thatcher at 10 Downing Street, while the Commission sent one to Alick Buchanan-Smith, Minister of State for Agriculture, deploring the non-agreement with the other countries on herring quotas and saying they viewed the whole situation with alarm. The Council delegated the Convener and Councillor Murdo Macdonald, Chairman of the Joint Committee to be part of a high level lobby of Government ministers on 17th February 1982 organised by Grampian Regional Council.

No longer a herring port!

It was January 1983 before six long years of negotiations ended in agreement on the Common Fisheries Policy and the Government issued a consultative document asking for the fishing industry's views on how best to re-structure the fleet to take advantage of the fishing opportunities now available to them. The Commission submitted that in order to preserve fishing communities, particularly on the west coast of Scotland, local management schemes were necessary. They did not favour laying up premia where the majority of the funding would go to the larger vessels and be of little or no use to the smaller fleets; but they urged that quotas should be on based on the number of crew, which although not perfect, was the best method available.

When it was announced in 1982 that Stornoway had been removed from the list of designated herring ports, Murdoch Macleod asked the Department of Agriculture and Fisheries for Scotland for the reasons. He pointed out that prior to the closure of the Minch, the port had been heavily committed to the klondyking operation and said it seemed unfair that Stornoway, whose very existence as a port arose from the Minch herring fishery, should be deprived of the facility for herring landings which they had held for hundreds of years.

The reply: the Department in Edinburgh denied that there was any attempt to direct the herring fishing away from Stornoway; indeed the Department would welcome any klondyking operation at the port. Unfortunately, he said, there seemed to be no substantial market for human consumption at Stornoway and since the fish factory was receiving so much poor quality fish rejected at other ports, it was full to capacity and unable to process the good quality catches, which were then having to be dumped. The Department did not see much sense in this situation. He stressed, however, that if a catch failed to find a human consumption outlet, then the boats would be free to go to Stornoway, with permission from the Fishery Inspector.

This did not satisfy the Commission who requested answers to specific questions posed by the General Manager. Mr Stewart wrote again with these, informing them that the proportion of herring which went for fish meal at Ullapool and Stornoway was 15% and 97% respectively. He also admitted that the opening of the herring fishery in 1981 following the ban had been more hurried than the Department would have wished but denied that this had lost the klondykers to Stornoway. Mr Stewart again stressed that their decision had been purely a response to the 1982 situation. The Department appreciated Mr Macleod's point that it would be difficult to attract klondykers to Stornoway unless it was a designated port, and therefore had decided that it should be so designated for 1983.

The system did not seem to work in practice, however. In September 1983 the General Manager reported that four Russian klondykers had come into Stornoway to take herring from the Norwegian fishing fleet. With the opening of the Minch herring fishery the Russians had wanted to stay in the harbour for that fishery and possibly also the mackerel season. Their agents had been advised that Stornoway was no longer a designated port for the transhipment of herring, and the Russians had left for international waters. The local Fishery Officer told an irate Murdoch Macleod that the Licences for the herring fishery did not show Stornoway and according to his superiors in Edinburgh, although it was a now designated port again, permission still had to be received from Ullapool. This state of affairs had lost the Commission an estimated £60,000 in harbour dues and the General Manager took this up with John Mackay, Minister of State at the Scottish Office when he visited Stornoway shortly afterwards.

The Department insisted that this incident had been a misunderstanding. The klondykers had left simply because the Norwegians could not supply the minimum daily catch for the vessels, not for any other reason. The Scottish Secretary, George Younger, replying to the Member of Parliament Donald Stewart, backed the officials. The Commission agreed to make more representations to him on the lack of involvement of Stornoway fishing interests in the consultation process. They also objected strongly to any suggestion that the District Inspector at Ullapool should be referred to for landing mackerel or herring at Stornoway.

The official line was that the industry had been consulted and had itself taken the decisions on how best to manage the herring fishery and the landing ports. The island fishermen maintained that they had no representation on the bodies which had been consulted. In the summer they decided to withdraw from the Mallaig and North West Fishermen's Association and form their own. The Commission informed the Department that the new organisation which should be consulted with regard to fisheries affecting the Minch was the Western Isles Fishermen's Association.

In an attempt to attract some of the klondyking business back to the port, the Harbourmaster and the Harbour Collector both contacted the Ullapool agents, while the General Manager, in Hull for a British Ports Association meeting, took the opportunity to meet a representative of J Marr & Son Ltd, the trawler owners, who also had interests in the west coast operations. The efforts had some success. Two Dutch klondykers arrived in January 1985 to take mackerel from Irish vessels, fishing outside the 12 mile limit and happy to move from Ullapool to save time.

In June 1985 Donald Stewart wrote to say that he had raised the matter with Lord Grey of Contin, the Minister of State and he did not see what could be done to change the Government's attitude. Lord Grey stated that Stornoway would not become a designated port again until a demand by the human consumption market there justified it. Further, if vessels were allowed to land, without prior authority from the Fisheries Inspector, there was a real danger of fishing directly for industrial purposes to take place, and this was in contravention of EEC regulations. The arrangements would have to continue.

A New Terminal Needed

At the end of 1985 the Commission took a decision to close off the piers to all but harbour users. The increase in traffic and the proximity of the quays to the town made it increasingly difficult to control and indiscriminate parking added to the problem. Gates were put in, closing off No 2 Pier and Cromwell Street Quay and the General Manager submitted a report on traffic control arrangements at the ferry terminal. This sparked considerable discussion and a site visit, after which the Council offered the assistance of their Engineering Department in devising a pattern for control at ferry times.

A meeting to consider the whole problem was held in Amity House on 22nd June 1988, attended by the Managing Director and the local manager of Caledonian MacBrayne Ltd, the Director and Assistant Director of Engineering Services at the Council, and the Inspector of Police. In the short term the meeting agreed on some measures that could be taken such as access by disabled, the marshalling of taxis, and the possible streaming of arriving passengers through the gap between Maritime Buildings and the Transit Shed.

A consensus emerged that the problem at No 1 Pier could never be solved until the Commission had carried out a traffic management exercise, taken what was essentially a haulage yard off the pier, had arranged for the segregation of arriving and departing passengers, and had found more effective ways of controlling the movement of heavy goods traffic in the marshalling area. The Highlands and Islands Development Board having been approached said they were aware of the problems and were willing to look at financing a Traffic Management Survey.

Commissioner Malcolm Macfarlane called for MacBrayne's to provide a fast, new custom-built vessel for the run, pointing out that the *Suilven* had no lift and disabled passengers had to be assisted, if not carried, from the car deck to the passenger decks. At the meetings on traffic management he stressed the need for a proper terminal for passengers, as well as a covered gangway, facilities, he said, which were long overdue.

In November 1989 a meeting between all the interested parties, plus the Highlands and Islands Development Board, agreed that Consultants should be engaged to carry out a study, the specification for which was drawn up by Murdoch Macleod and Colin Macaulay, the Comhairle's Director of Engineering. This asked for a report on the feasibility of constructing a new Ferry Terminal, either in the Newton Basin or at Glumaig Bay 'where a Fabrication Yard for the oil industry is situated but presently unworked'; the detailed study of existing operations, giving proposals for any improvements; estimated costs for possible development; and the identification of sources of funding.

Five firms tendered for the study work and were interviewed, with the lowest price being that of Blyth and Blyth, Inverness. There were two reports to be produced, one on Ferry Terminal Facilities, the other on Facilities at Stornoway Harbour. The Council contributed to the cost and the Highland Board offered over £5000 in grant.

At the beginning of the 1990s the Commission's financial situation was not encouraging. Although the accounts had shown a profit, revenue was down, attributed to the decline of the fishing industry, the slump in the Harris Tweed market, and the closure of the oil fabrication yard. The Commission had incurred expenditure on renewing the Harbour lighting, installing a new weighbridge and commissioning the ferry terminal study. Still in the pipeline were the delayed repairs to No 1 Pier fendering, new windows for Amity House, and the purchase of a pilot boat, if a suitable vessel could be found.

Frank Apps, giving his report, said that if the budgeted expenditure had been carried out, there would have been a loss of £15,000. He did not see a recovery for the Commission's total income in the present financial climate. The foul weather had disrupted the winter's klondyking operations, the cranes were more of a liability than an asset, there had been a decrease in the number of lorries using the weighbridge because of the dispensations for unladen vehicles; and there was no way the capital cost of the weighbridge could be recovered. The one bright spot, he said, was the slipway, adding that given all these factors the accounts were reasonable.

Reginald F Apps had been a member of the Commission for ten years, and for seven of those chaired the Finance Committee. His death in October 1990 was a shock, removing one of the Commissioners' steady hands and financial experts. A meteorologist by profession, he had worked in Palestine, the West Indies and Hong Kong before returning to Stornoway with his family. In Hong Kong he had been Director of the Royal Hong Kong Observatory. As the Chairman, Malcolm Gillies, said, all the bodies he worked with had been immeasurably better off for his presence among them. It was sad that he did not live to see the new terminal in operation.

In February 1991 Blyth and Blyth submitted their findings and recommendations, coming down in favour of the Newton Basin and presenting the various options to the Commission, Council and Highland Board. The discussion ranged on a number of points, including the need for a new linkspan; the need for extensive repairs to No 1 if it was to be developed for the terminal; and the requirements of the new ferry scheduled to take over the run in 1994. It was decided to take things further by preparing a traffic management scheme and looking into the question of funding.

This was to be the major project for the 1990s and the major achievement of the Stornoway Pier and Harbour Commission's last decade. At their meeting in January 1992 the Commissioners settled on the option to build a new terminal in the Newton Basin and started the ball rolling. An Economic Appraisal was carried out by Pieda plc, with a view to presenting a case for financial assistance; a Port Risk Assessment Study was commissioned from DNV Technica and a Working Group of Commissioners set up to deal with the whole subject. An in-depth study of the deterioration in the existing pier was also commissioned and carried out by Babtie, Shaw & Morton, Glasgow, Consulting Engineers.

By the beginning of 1995, the Commission knew they had to find £7.6m to blast the seabed in the Newton Basin, infill a huge section of the harbour there, construct a pier, a linkspan, a marshalling area for traffic and a terminal building. They had secured 40% of that from the European Regional Development Fund. The Scottish Office would fund 75% of the remainder, which left a £1.19m balance. The Commission decided to fund this by arranging a loan from the Public Works Loan Board repayable over 25 years.

A new terminal built

The project was managed by Babtie's who sent their own Resident Engineer, Eric Macaulay, to oversee the works for the next two years and the Commission's Sub-Committee met every four weeks to monitor progress. Comhairle nan Eilean Siar carried out road improvements and introduced roundabouts to control the flow of traffic to and from the new facility and worked closely with the Commission over the months.

There were seven separate stages of the work, starting with dredging of the seabed by Westminster Dredging on 1st May 1995. Two months later Messrs Fyfe began the land reclamation of the site, infilling with material from the local quarries. This took five months and, by the end of the first year, Balfour Beatty were beginning the construction of the new pier.

Meanwhile, over in Arnish, the Lewis Offshore Yard was working on the new 40m linkspan, engineered by William Arrol's. The pier and the linkspan took the best part of a year to complete, the linkspan being swung into position on 18th June 1996, and the pier contract finishing one month later.

The new terminal building, based on the shape of the old octagonal Fish Market, was designed by the Inverness architects, Wylie Shanks, and the construction carried out by the building firm of R J Macleod Ltd, Dingwall, starting on 6th May 1996 and finishing just a week before the opening date of 27th April 1997. Fyfe's carried out the external works contract on the marshalling area, roadways and pavements over the winter of 1996-97, as well as the final part of the project, the demolition of the Transit Shed on No 1 Pier, from August 1997 to March 1998.

The building had offices for Caledonian MacBrayne and luggage reception and reclaim areas for William Macdonald & Son, the bus operators. A large viewing area with seating allowed the public to wait in comfort at ferry arrival and departure times and a cafeteria area, toilets and baby changing facilities were also provided. The remaining space, which could accommodate two small retail units, was eventually taken over by Caledonian MacBrayne as an extension to their offices. A covered walkway from the building to the ferry gangway completed the building.

The crowning touch to the new building was the installation of a custom-built weather vane. The idea of commissioning Fred Dibnah, television personality and steeplejack extraordinaire, to fashion the weather vane had been John Maclennan's. Mr Dibnah had created a stainless steel vane in the shape of a traditional *sgoth Nis* and three of the main users of the terminal, Caledonian MacBrayne, and the hauliers, Colin Macaskill and D R Macleod willingly sponsored the work. The inimitable Fred made the journey to Stornoway in May 1997 to fix the weathervane personally on the roof of the new building.

Another addition to the building was the original drinking fountain from the old Fish Market, donated by Mrs Jean Mackintosh, who had rescued it from the rubble over twenty years earlier. It had lain undisturbed in her barn at Achmore until 1997. 'It is now in a perfect place,' she said. 'It never really belonged to me, it belongs to the people of Stornoway.'

The new facility opened on 27th April 1997 and the first passenger and first car driver to disembark from the *Isle of Lewis* that evening were presented with Stornoway Pier & Harbour Commission plaques to mark the occasion. It was officially opened on 14th May 1997 by the grand old man of the Lewis scene, James Shaw Grant, on one of his last public appearances. As ever, his speech was masterful, ranging back to his childhood days and far beyond, and highlighting some of the changes in the life of the town since then. He ended with the words:

> 'Stornoway still flourishes in the only way it can – on the enterprise of the inhabitants of the island of Lewis. In your name I would like to thank all those whose foresight and hard work has made possible this splendid ferry terminal – our gateway to the wider world.'

Income and Expenditure

The ferry terminal project had given the Commissioners and the General Manager some sleepless nights before its completion. Its financing had been a huge undertaking at the time and until the European grant had been confirmed the whole project had hung in the balance. Cash flow over the three years of the construction had required strict monitoring.

Throughout the decade the Commissioners took several measures to encourage more trade at the port. In 1991 they joined the newly constituted Cruise Europe group, set up to market Northern European

Fred Dibnah, designed and personally fitted the weather vane on the roof of the new terminal building

ports as suitable destinations for cruise liner tours. In June 1996 they invited representatives from the Council, the Western Isles Enterprise, the Tourist Board and the Chamber of Commerce to a meeting, from which a Co-ordinating Group was formed. A marketing study was commissioned on how best to tackle the promotion of Stornoway as a cruise liner destination, and on one or two occasions, delegates were sent to attend the Seatrade Conference in Miami as part of the Cruise Scotland team. The number of liners calling at the port now averages around 17 in the months between May and September.

A concerted effort was made in the 1990s by the local public bodies to attract some of the oil business to the port. The Western Isles Oil Group was a promotional group which for over a decade worked to promote the islands to the oil industry. The Harbour Commission was one of the members and regularly sent representatives to help man the stand at the Offshore Europe Exhibitions in Aberdeen.

As well as looking to the oil industry for business, the Commissioners looked into the possibility of attracting European fishing vessels to the port. They set up a working party in 1992 and delegated representatives to visit trawler owners and associations in France and Spain with a view to encouraging them to land their catches at Stornoway for transhipment to the continent.

The General Manager and the Deputy Harbourmaster visited Concarneau in Brittany in early December 1992, calling on six French fishing companies over two days. One had already invested heavily in Lochinver and was unlikely to move operations, and the others, while impressed with the facilities available at Stornoway, all expressed reservations about transport costs.

Caledonian MacBrayne lodged an objection in March 1996 with the Secretary of State for Scotland against the amount of dues charged by the Commission on the new vessel *Isle of Lewis*, which had come on the run in 1995. They claimed among other things that the amount of the increase was excessive and that the company, while using only 50% of the Commission's turnover, was being asked to pay the whole of the harbour expenditure.

The Scottish Secretary came down in favour of the Harbour Commission, after over a year of submissions, questions and information, winging its way between St Andrew's House and the Harbour Office. John Maclennan's case for the charges was accepted by the Secretary of State in September 1997, in which Donald Dewar stated that the Commission's income

> '. . . is sufficient only to meet the Commission's expenditure and to make prudent provision for future repairs which is consistent with the Commission's financial regime.'

Besides the ferry terminal the Commissioners carried out several other major projects over their last decade. The EC directive on Fish Hygiene Regulations meant an upgrading of the Fish Market involving insulating the building at a cost of over £100,000; a pilot boat was purchased in 1993 for £47,000 now that the Commission had become a Competent Harbour Authority for providing pilotage services; and pontoon moorings for visiting yachts were installed at Cromwell Street Quay in 1995,at a cost around £100,000, also assisted by Western Isles Enterprise.

Slipway improvements costing £469,000 were undertaken in 1998, with funding from the ERDF and WIE. The contract was awarded to John M Henderson's, Aberdeen, who as well as the Consulting Engineers, Arch Henderson and Partners, had worked on the original construction over forty years before. The net sheds on Cromwell Street Quay, which had become run down and a magnet for vandals, were demolished in 2003, replaced by 72 plastic net bins purchased from Norway, and sited on Station 29, North Beach Quay. The £98,000 costs were also part funded by the FIFG scheme and Western Isles Enterprise.

A walk round the pier

The changing nature of society in the last years of the twentieth century forced the Commission into taking unpopular measures in its own interests. Apart from Health and Safety legislation, various Acts of Parliament and Regulations were passed over the years, laying responsibility on port authorities for security and for protection of the environment. In 1998 Captain Mackenzie reported on the need to fence off the whole port area at a cost of £60,000 in order to meet the Commission's legal responsibilities for safety and security. This was never going to be a popular decision, since access to 'the harbour' had been part and parcel of the community's daily life for centuries. Even among the Commissioners themselves there was some hesitation but the Harbourmaster made the point that the intention was to control, not prohibit, public access.

The erection of dwarf walls topped by ornamental railings around the car park areas in the Inner Harbour are now accepted as part of the town scene, echoing the railings in the town square a century earlier. The fences around the entrance to No 1 Pier and Amity House are no longer seen as a restriction but a control and the public is still free to walk around the area, unless there are port operations in progress.

Indeed sometimes the public were positively welcomed around the piers as in the summer of 1996, when the Transit Shed on No 1 Pier was transformed into a theatre. The Commission agreed to the use of the Transit Shed with seating for 250 people being situated in the larger end of the building. The two performances of the community play 'Portrona' written by Norman Macdonald from Tong and produced and directed by Muriel Macleod from Stornoway were packed out. The story centred on events in Stornoway when the town was in its heyday as a herring port and was a huge success.

In May of the following year another unique event took place at No 1 Pier when Britain's only touring theatre ship *Fitzcaraldo* spent three days in port. The public enjoyed the experience of going aboard to watch 'Gulliver's Travels' performed for family audiences by the Walk the Plank Theatre Company based in Liverpool.

Twenty-one vessels participating in the 1999 Cutty Sark Tall Ships Race called at Stornoway at the beginning of August on their way from Greenock to Lerwick. Their stop at the port was organised by the new volunteer organisation 'Sail Hebrides' set up in 1997 to encourage and promote sailing among the young of the island. A welcoming ceremony was held at the new Terminal Building.

The RNLI's naming ceremony for their new boat *Tom Sanderson* was also held there two days later while on the same day passengers from the 37,845 ton cruise liner *Royal Viking Sun* anchored in the harbour came ashore to sightsee. The sun shone for all these events, which ended with the annual Fish Festival on the Saturday afternoon, run by the Royal National Mission to Deep Sea Fishermen. Sail Hebrides went on to run an annual Maritime Festival at the port, which concentrates on small boats and traditional sailing craft, as well as introducing youngsters to the delights of sailing.

The Pipe Band were allowed for a time in the 1990s to practise in the Dockers' Rest Room on South Beach Quay, by this time very rarely used, and also to store some of their equipment in the lockers there, until the completion of their own premises. This part of the No 3 South Beach Street was eventually rented to Stornoway Shipping Services for their ship's agency.

In the 1990s the Fishermen's Co-operative found themselves unable to continue with their plans for the Cromwell Street Quay building and accordingly demolished it. Now that the edifice which had caused so much controversy was gone, the Commissioners considered how best to deal with the vacant space and began by lowering the remaining wall and restoring the views across to the Castle grounds. In 1998 they agreed in principle to provide facilities for yachtsmen comprising toilet, showers and laundry facilities at the Old Lifeboat Shed opposite the pontoon facility, and also a winter hard standing for yachts on the vacant site. The hardstand exists but the yachting facilities have yet to be put into action.

Among the other changes around the piers were the demolition of the old bus station on Cromwell Street Quay; a new shore facility for the RNLI with a new berth for the lifeboat directly opposite; the closure, amid strong protest from the community, of the Stornoway Customs Office on Quay Street; and the sale of the Old Sail Loft and Commercial Hotel to the Highland Buildings Preservation Trust for £1. Requiring large sums of money to repair, let alone restore it, the Commission could not access funding to deal with the problem, but under the hands of the Trust it has been transformed. The Loft is now a block of three houses, while the Commercial Hotel has six flats, all lovingly restored and crafted to make the best use of one of the oldest buildings in Stornoway.

At the other end of the harbour, the Coastguard's headquarters building, Clan McQuarrie House, named after one of their most notable rescues, stands at Battery Point overlooking the harbour entrance. The plot of land was purchased from the Commission in 1992. The Commission also sold their interest in the site of the Sailor's Home on North Beach to Comhairle nan Eilean in 2000.

The Stornoway Amenity Trust were granted permission to carry out improvements at Newton and create a garden next to the Ferry Terminal entrance, as well as providing hanging baskets round the terminal area. It was a far cry from the days when Norrie Maciver had vowed that there would be no rhododendrons on the pier while he was around.

Not another listed building

The Conservative Government's legislation in 1991 included provisions for Harbour Authorities to take over navigational lights within their area. The Commission, faced with having to take over the maintenance of the Arnish Lighthouse and the buoy, put up a fight against this. They lobbied members of the House of Lords and pressured the MP but the Bill went through.

In October 1991 they received the Lighthouse Board's proposals as to which lights would be transferred to them. These were the Arnish Lighthouse, the lighted buoy at Arnish Point, the unlighted beacon at Seid Rocks and the unlighted beacon at Biastan Thiulm. The Commissioners resisted this, on the grounds that the lighthouse was not solely for the use of vessels entering Stornoway Harbour and that the financial burden was unacceptable when they had no way of recompense. The Secretary of State for Transport accepted the argument, noting that the 18 mile range of the Arnish Lighthouse would mean there was some degree of usage for general navigation, and that this would be difficult to quantify.

In 1998 the Board returned to the subject of the Arnish Lighthouse, proposing to decrease the range of the light to 9 miles, which would then effectively do away with the case for its being used by vessels in the Minch. Again the Commissioners resisted this idea, pointing out that there was increased tanker traffic in the Minch for whom the full range of the light was necessary. They reserved the right to refuse the cast iron lighthouse, built in 1852 by the Stevenson family and a Category B listed building.

The Northern Lighthouse Board's offer to maintain the lighted buoy and the two unlighted beacons on an annual contract basis was accepted by the Commission and in 2000 they were informed that the Board would refurbish the lighthouse to the tune of £48,000, as well as convert the light to solar electricity before handing it over.

The old and the new

In 1999 two wooden benches were purchased for the terminal building and inscribed to the memory of two men of a previous generation. Malcolm Macdonald, better known as 'Calla Dol', served on the Stornoway lifeboat for forty years, twenty of those as coxswain. During that period, says the inscription, 188 lives were saved. Angus 'Ganga' Slater, was one of Stornoway's best known and fondly remembered characters. He spent his life around the harbour; grew up in the crowded streets close by the quays where he later worked as a fish cadger and a fishermen. The wording on his bench was suggested by his family and says that it is erected by the Commission 'in memory of Ganga Slater 1923-1976 in recognition of the 19 lives he saved from this harbour – a true cove'.

In the summer of 1991 Murdoch Macleod received an unexpected letter from a lady in Grimsby who wrote that she believed she was in possession of stolen property. Her father had been a scrap dealer and had purchased a bronze plaque from a firm near the Grimsby docks sometime in the late 1970s. Realising it commemorated a royal visit and thinking it related to Grimsby, he asked his local council of they wanted to keep it, but nothing was done and the plaque lay in his garage until his death in 1985. An antique dealer had told his daughter that he had a prospective buyer who collected Royal memorabilia and bore it away to show him. He in turn showed it to a reporter on the *Grimsby Evening Telegraph* who knew a good story when he saw one and ran an article on the plaque and its origin – which was when the lady found out that she was in possession of the bronze plaque that was attached to King Edward Wharf commemorating the 1956 visit of the Royal Family. The plaque was safely returned to the harbour shortly afterwards

Congratulations were conveyed to boatbuilder John Murdo Macleod in December 1994 on the completion of his *sgoth Nis* project. The beautiful *An Sulaire*, which he had lovingly created in the style his father and grandfather had done before him, was a magnificent achievement, the Commissioners told him, and had brought great credit to himself, to Ness and to the island as a whole.

Two well-known Stornoway exiles, Eddie Glennie and Johnny

'Cannutt' Macdonald, on holiday from Vancouver, paid a visit to Amity House in September 1997 to present a reproduction painting of the SS *Metagama* and a porcelain plate depicting Scottish emigrants arriving in America. The painting was hung in the new ferry terminal while the plate remains in the Board Room of Amity House. In 2002 Frank Thompson gifted the Commission with a framed 1775 chart of Lewis by Mackenzie for the Boardroom and a painting by Laura Symes, depicting the old Fish Mart on a busy day was hung there as well in 2003.

The Stornoway Historical Society suggested that their 1995 annual Exhibition should have as its theme the Stornoway Pier and Harbour Commission which would be then in its 130th year. This was mounted in the Town Hall and displayed a wealth of information from the Commission's records – Minute Books, Letter Books, Cash Books, and Arrivals and Sailings Ledgers, dating back to the Commission's inception, as well as some legal documents from the eighteenth century. The need to preserve these records was becoming more and more urgent, given their unique and valuable contribution to the history of the island, as well as the port itself.

The Commission decided in 1998 to have the oldest of the ledgers restored and microfilmed to preserve the record. Some had been badly smoke and water damaged in the Town Hall fire from which Archibald Munro and Major Bain had rescued them in 1918. The Scottish Record Office gave a grant of £2,000 over two years towards having the volumes rebound by Valentine's Bookbinders in Larbert, while the Museum Committee contributed £200 over five years towards the project. With the collaboration of Western Isles Libraries, the Commission had the volumes microfilmed by the Scottish Newspaper Microfilming Unit. A copy of the microfilm is available for viewing in Stornoway Public Library and the volumes themselves can be accessed in Amity House by prior arrangement.

In 1995 the Commission fully computerised its accounting procedures and record systems. Before deciding on how to proceed Murdoch Macleod wrote to 16 port authorities of a comparable operation to Stornoway asking for information about their systems. Among the detailed and earnest replies received, came the following:

> Dear Murdo,
> Received your wee note re computer accounting systems. I would advise that we are still using the of 10st 11lb model installed some years ago. It huffs and puffs occasionally but seems to function reasonably well when it is maintained with regular modest lubrications of good quality blends. It also operates during our not infrequent power cuts.
>
> There have been no complaints from the Trustees, the Auditors,

> Inland Revenue, Scottish Office or Brussels audits, so it is therefore assumed to be effective.
>
> Felicitations and my kindest regards to all at Stornoway Harbour,
>
> Alan

The writer, for all his wry words, was well in control of his harbour's affairs.

Into the new century

Murdoch Macleod retired in August 1996 after almost thirty years with the Commission, 21 of these as General Manager, Secretary and Treasurer. The new General Manager was John J Maclennan, the Depute Secretary and Collector of Dues. The Collector's post was offered to Catherine Mackay, the Commission's Administrative Assistant. At Murdoch Macleod's last meeting on 21st August 1996, the Chairman, Sandy Matheson, marked the occasion by reminding the Commissioners that Mr Macleod had been at the centre of community life for over forty years, the best example of a dedicated and diligent servant of his community and its people. Remarking on the various capital developments that had taken place during his time in post, the Chairman added that raising the funding for the new ferry terminal had been a *tour de force* and the Commission and the community were indebted to Murdoch Macleod for his immense contribution to the project.

In common with businesses and organisations all over the world the Commission prepared itself for the turn of the twentieth century into the twenty first. There was a fear abroad that all kinds of equipment and systems might fail if computers and timers were not able to adjust to the change from 1999 to 2000 without going back to the beginning. As well as that, celebrations including a Millenium Fireworks Display, would be held well into the early hours of January 1st. As well as normal duty watchkeeper, the General Manager, the Harbourmaster, his Deputy, the Security Officer and one Port Operative were detailed to be on stand-by for call-out between 8 pm on NewYear's eve until 4 am in the morning. Extra fuel, an electricity generator and extra VHF radios were laid in but the night passed without incident.

The Slipway Superintendent, Roderick Macritchie retired in 2000 after almost a quarter of a century. His successor was the Slipway Assistant, Finlay Campbell from Stornoway. Angus Mackenzie retired on 8th April 2004. His deputy, Angus Macleod, became the last Harbour-master of the old Commission for its remaining three weeks and first Harbourmaster of the new Port Authority from 1st May. The

Commission's Security Officer took on the combined role of Deputy Harbourmaster, Traffic Control and Security Officer with the new Port Authority over the same periods

Another beginning

The new century began with a major change for Stornoway Pier & Harbour Commission. The new Government had decided in to make a radical change in the way public trust ports were administered, following on from the findings of their Trust Ports Review undertaken in 1997. Their guidelines 'Modernising Trust Ports – A Guide to Good Governance' were issued on 10th January 2000 and had among its key points that harbour boards should have between 8-12 members, serving a maximum of three years renewable twice, and appointed by a system of open competition and advertisement. They were not to be restricted to councillors and at least one member of the board should be an executive member, such as the Chief Executive or Operational Manager.

There were to be eight guiding principles for the conduct of the members. These were based on the Cabinet Office Code of Practice for Board Members of advisory public bodies and were independence, accountability, openness, selflessness, integrity, objectivity, honesty and leadership. As well as the management of the port, members would have to be active in seeking business opportunities and working with other bodies.

A Special Meeting was held on 25th May 2000 to study the matter and decide on what measures to take to comply with the Guide to Good Governance. The Commissioners had a long and varied discussion and decided to enlist outside help with the required Implementation Plan and Harbour Revision Order that they were obliged to produce. A Working Group, chaired by Commissioner Kenneth M Macdonald was set up to take the matter forward.

The resulting plan for implementation put forward by the Working Group was that the new Board would still consist of 10 members – one would be the Chief Executive, two nominated by Comhairle nan Eilean Siar, one nominated by Stornoway Trust, and two nominated by the Commission. The remaining four members would be elected by the Lewis voters. The restriction of voting to duespayers only was abolished as was the right of the Sheriff Principal to appoint a representative.

All candidates, including those standing for election, had to first meet the National Standards of fitness for purpose, openness, and accountability, and meet the requirements of a job description.

The Scottish Office accepted the plan and the Commission went on to promote a new Harbour Revision Order to change the constitution. They also agreed to change the name to Stornoway Port Authority and the General Manager's title to Chief Executive and to extend the Harbour Limits

The last Commissioners to serve under the old constitution and name were Angus Campbell, Keith Dodson, Kenneth N Macdonald, Norman L Macdonald, Malcolm Macfarlane, Donald H M Maciver, George Maciver, Iain A Macleod, Sandy Matheson and Ruaridh Morrison The final meeting of the Commission was held on 28th April 2004. As the Chairman Iain Macleod closed the business, he said:

> 'The name Stornoway Pier & Harbour Commission ceases to exist on Friday 30th April 2004 and this organisation will be re-named the Stornoway Port Authority. . . These are exciting times in the life of this 139 year old institution and I for one look forward to the immense challenge ahead. . .'

THE END

Stornoway Pier and Harbour Commission
Names 1865 to 2004

Chairmen

Sir James Matheson	1865-79
Lady Matheson	1879-96
Donald Matheson	1896-99
Major Duncan Matheson	1896-1918
Viscount Leverhulme	1918-24
Hugh Macleod	1924-26
Murdo Maclean	1927-30
Norman Maciver	1930-32
Roderick Smith	1932-36
William J Tolmie	1936-39
David Tolmie	1939-40
Cpt Thomas C Macdonald	1940-47
Norman Maciver	1947-60
Edwin Aldred	1960-66
Donald J Stewart	1966-69
Alasdair Matheson	1969-71
Alexander Matheson	1971-72
John Macleod	1973-83
Malcolm Gillies	1983-91
Alexander Matheson	1991-2001
Iain A Macleod	2001-04

Officials

Secretary/Treasurers/General Managers

John Ross	1865-1889
John & William Ross	1889-1891
John Ross	1891-1896
Archibald Munro	1896-1931
Alasdair Macleod Secy	1931-1933
Alasdair Macleod	1933-1968
Murdoch Macleod	1968-1996
John J Maclennan	1996-2004

Harbour Masters

Peter Macfarlane	1864-1865
William Lees	1865-1887
Thomas Morrison	1887-1913
Malcolm Maclean	1913-1919
George Watt	1919-1924
Alexander Finlayson	1924-1960
Alexander Mackay	1960-1978
Angus M Mackenzie	1978- Apr 2004
Angus M Macleod	April-May 2004

Harbour Collectors

William Lees	1865-1896
Joseph Coupar	1896-1897
Alasdair R Campbell	1897-1912
Angus Macleod	1912-1922
William Grant	1922-1941
William J Macdermid	1941
Angus Macdonald	1941-1959
Donald Murray	1959-1982
John J Maclennan	1983-1996
Catherine Mackay	1996-2004

Commissioners

Sir James Matheson	Proprietor, Lews Castle	1865-79
Donald Munro	Chamberlain c/o Carn House, Sth Beach	1865-75
Kenneth Smith	California, 12 Lewis Street	1865-1900
Norman Maciver	Shipowner, 12 South Beach	1865-76
Daniel Macalister	Shipowner	1865-74
Capt Donald Mackenzie	Shipowner, Kenneth Street	1865-1877
Kenneth Mackenzie	Bank Agent, National Bank, 17 N Beach	1865-90
James Mackenzie	Merchant, 21 Cromwell Street	1865-72
Alexander Morrison Jnr	Fishcurer, Kenneth Street	1872-78
Roderick Geo Mackenzie	Shipbroker, Kenneth Street	1873-83
Matthew Russell	Merchant, 4 Cromwell Street	1874-89
William Mackay	Chamberlain, Carn House, Sth Beach	1875-89
Roderick Daniel Mackenzie	Merchant, 27 Cromwell Street	1877-78
Aeneas Mackay Mackenzie	Shipowner, 25 South Beach	1878-81, 1891-93, 1899-1904

Angus Smith

Angus Smith

The last Stornoway Pier and Harbour Commissioners and officials outside Amity House:
Back Row: Captain Angus M Mackenzie, Harbourmaster; Keith Dodson; Ruaridh D Morrison; Norman L Macdonald; John J Maclennan, Chief Executive; Kenneth N Macdonald; Captain Angus M Macleod, Deputy Harbourmaster
Front Row: Angus Campbell; Donald H.M. Maciver; Malcolm Macfarlane, Vice-Chairman; Iain A. Macleod, Chairman; Sandy Matheson; Catherine Mackay, Collector of Dues. Insert: Captain George Maciver

Malcolm Macleod	Vintner, 39 Kenneth St	1878-81
Lady Mary Jane Matheson	Proprietor, Lews Castle	1879-96
John Mackenzie	Merchant, 21 Kenneth Street	1879-90
Murdo Macleod	Bank Agent, South Beach	1881-91
Thomas Morrison	Ret'd Sea Captain, 49 Francis Street	1883-85
Murdo Morison	Merchant/Fishcurer, Kenneth Street	1885-93
John Munro	Joiner?, 18 South Beach?	1886
William Maclennan	Chamberlain, Carn House, South Beach	1889-91
John Mackenzie	Fishcurer, South Beach	1889-1920
John Mackenzie Morison	Steam boat agent, 2 Bayhead Street	1891-1912
William Gillanders	Chamberlain, Carn House, South Beach	1891-92
John Norrie Anderson	Shipowner, Francis St, then Plymville	1892-1920
Peter Macleod	Merchant, 19 South Beach	1893-95, 1897-1909, 1920
Ossian Robertson	Banker, 17 North Beach	1893-1926
William George Russell	Merchant, later 7 Matheson Road	1893-97
John Mackenzie	Shipbroker	1893-97
Duncan Maciver	Fishcurer Church St	1893-94, 1912-21
Donald Smith	Fishcurer, Francis St, then Springfield	1893-1916
Murdo Maclean	Merchant, 24 Lewis Street	1895-96, 1901-02, 1904-30
Charles Orrock	Chamberlain, Carn House, South Beach	1892-1920
Donald Matheson	Proprietor, Lews Castle	1896-99
Duncan Matheson	Proprietor, Lews Castle	1899-1918
John Macleod	Merchant, 55 Kenneth St	1896-1904
Angus Bain	Fish Salesman, Gowancroft	1898-1901
Robert Hogg	Hotelier, Lewis Hotel	1902-03
Kenneth Maclennan	Merchant, 7 North Beach	1903-09
Donald Maciver	Merchant, 5 Bayhead Street	1904-07,1913-17
Kenneth Mackenzie	Coal Merchant, Fernlea, 9 Matheson Rd	1907-27
John Macritchie Morrison	Fish Salesman, Berneholme, Goathill Rd	1907-1912, 1921-32
William A Ross	Solicitor, Wandene, Goathil Crescent	1909-15
Roderick Smith	Chemist, 32 James Street	1915-63
Alexander Maclennan	Merchant, 53 Lewis Street	1916-27
Angus Mackenzie	Flesher, 26 James Street	1917-21
William Hesketh Lever	Proprietor, Lews Castle	1918-20
George Stewart	Merchant, 3 Francis Street	1920-24
Col Lindsay		1920-21
Louis Bain	Fishcurer, 25 Lewis Street	1921-34
Archibald Macphail Fletcher	Chamberlain, Lews Castle	1920-24
Norman Maciver	Fishcurer, Barvas Lodge, Church St	1922-62
Murdo Maclean	Fishcurer	1924-30
Angus Cameron	Bank Agent, 17 North Beach	1926-27
Hugh Macleod	Chamberlain, 53 Kenneth Street	1924-27
Hugh Mackay Matheson	Merchant, Willowglen House	1927-31
Edwin Aldred	Chamberlain, 32 Scotland St	1927-66

Roderick Adam Morrison	Merchant, 34 Scotland St	1927-32
Robert Steven	Merchant, 38 Point St	1927-28,1930-40
William Mcnab	Bank Agent, 17 North Beach	1928-36
John Murdo Maclean	Coal Merchant, 39 Lewis St	1930-33
Alex John Mackenzie	Harris Tweed Mfcturer, Springfield Rd	1931-71
William Mackenzie	Docker/T7G Union Sec, 22 South Beach	1931-32,1936-38
William John Tolmie	Chemist, 49 Matheson Road	1932-39
Julia Fraser	The Old House, Lewis St	1932-37, 1940-46
Donald John Macdonald	Shipping Agent, 19 Lewis St	1933-50, 1952-60
Ossian Robertson Macaskill	Master Blacksmith, 53 Seaforth Road	1934-36
Donald J Maciver	Bank Agent, 17 North Beach	1936-52
Donald Gunn	Retired Teacher, 41 Matheson Road	1937-44
Ebenezer Mackenzie	Farmer, Melbost Farm	1938-39
Thomas Carr Macdonald	Local Fuel Overseer, Keith St	1938-48
David Tolmie	Commission Agent, 49 Matheson Road	1939-40
John S Maclean	Merchant, 82 Keith St	1940-50, 1952-53
John Poytress Newhall	Harris Tweed Mfcturer, Matheson Rd	1944-46
John Morrison	Timber Merchant, Goathill Crescent	1946-51
John Kennedy	Commission Agent, 72 Nicolson Rd	1946-52
Kenneth J Maclennan	Merchant/Oil, Rowantree, Goathill Rd	1948-53
John Murdo Bain	Fishmonger, 16 Ropework Road	1958-54
Robert Pringle	Tweed Mill Manager, Goathill Crescent	1950-56
Abraham Langley	Subpostmaster, 37 Nicolson Road	1951-68
Malcolm Smith	Hydro Board, 23 Francis St	1953-54
Robert W C McEwan	Shipping Agent, 25 North Beach	1953-55
William J Macdonald	Flesher, Knockgarry, Stornoway	1954-55, 1964-65, 1969-89
Albert Nicoll	Teacher, 26 Anderson Road	1954-58, 1971-78
Donald James Stewart	Harris Tweed Salesman, 5 Torquil Ter	1955-65, 1966-71
James R Macleod	Bank Manager, 17 North Beach	1955-62
Kenneth James Mackenzie	Harris Tweed Mfcturer, 5 Matheson Rd	1956-59
Ann Urquhart	1 Anderson Rd	1958-69
Charles Morrison	Merchant/Chandler, 6 Goathill Rd	1959-66
George Murray	28 Springfield Rd	1960-61
Roderick M Macleod	Tweed Designer, 42 Church Street	1960-64, 1966-71
Murdo Morrison	Secretary, 6 Matheson Rd	1960-61
Donald M Smith	Factor, Estate Offices	1962-64
James Macleod	Shipping Manager, 12 Goathill Rd	1962-66
Roderick Iain Maclean	Merchant, 12 Smith Avenue	1963-65
Alasdair Matheson	Hotelier, 1 Jamieson Dr	1964-73
John Macleod	Barber, 21 Macleod Rd	1965-83
James A Mackenzie	Travelling Salesman, 44 Westview Terr	1965-68
Callum R Morrison	Fish Processing Mgr, 5 Balmerino Dr	1965-68, 1971-72
James Maclennan	Building Contractor, 8 Smith Avenue	1966-68
John Macdonald	Company Director, 48 Point Street	1967-68
William G Lucas	Journalist, Ceol Mara, Springfield Rd	1968-74

Alexander Matheson	Pharmacist, 33 Newton Street	1968-2004
Kenneth Maclennan	Company Director, 16 Barony Square	1968-76
Donald Macleod	Company Director, 6 Barony Square	1968-71, 1973-80, 1984-92
James Macrae	Millworker, 8 Robertson Rd	1971-84
John M Morrison	Fishing Skipper, 2 Barony Square	1971-98
Alexander Bruce	Company Director, 29 Lewis St	1971-73
Alasdair Macrae Mackenzie	Harris Tweed Director, 12 Matheson Rd	1973-82
Murdo Macleod	Teacher, 31 Urquhart Gardens	1974-78
Malcolm Gillies	Accountant, 47 Matheson Rd	1975-93
Marie Alpine Macmillan	32 Matheson Road	1978-86
Reginald F Apps	Ret'd Civil Servant, 42 Barony Square	1980-90
Donald Macaskill	Fishing Skipper, 13 Olivers Brae	1980-95
Kenneth Macdonald	Ret'd Police Officer, 7 Barony Square	1978-80
Murdo Afrin	Chiropodist, 4a Sand Street	1982-94
Malcolm K Campbell	Fishing Skipper, 36 Westview Terrace	1983-86
Malcolm Macfarlane	Meteorologist, 4 Anderson Rd	1986-2004
William H Watson	Shipping Manager, 23 Lewis Street	1986-92
John Crichton	Crofter, 21 Swordale	1990-94
Donald Hugh Morrison Maciver	Sandwick Cottage, Millar Rd	1990-2004
Norman L Macdonald	Valasay, 11 Goathill Crescent	1992-2004
Iain Archibald Macleod	4 Steinish	1994-2004
Frank S Burns	22 Seaview Terrace	1994-1999
Alasdair Gillesbuig Macrae	Undertaker, 59 Keith St	1994-2003
George Maciver	Sea Captain, 17 Barony Square	1995-2004
Ruaridh D Morrison	Fishing Skipper, 13 Tolmie Terrace	1995-2004
Kenneth N Macdonald	Dentist, 5 Lewis St	1998-2004
Angus Campbell	8 Bulnacreig Street	1999-2004
Keith Dodson	34 Mackenzie Street	2003-04

Catherine Mackay was born in the village of Achmore and educated at the Nicolson Institute, Stornoway.

After leaving school Cathie joined the firm of Kenneth Mackenzie Ltd which was then one of the major producers of Harris Tweed in the Outer Hebrides.

In 1974 she entered into the employment of the Stornoway Pier & Harbour Commission as Collector's Clerk to the then Harbour Collector, Mr Donald Murray. In 1996 she was promoted to Harbour Collector, the first woman to hold this post since the inception of the Commission in 1865.

Cathie is unrivalled in her knowledge of the history of the Stornoway Pier & Harbour Commission and in addition to her duties as Harbour Collector she has set up an archive of the history of the Commission which will prove an invaluable record for future generations.

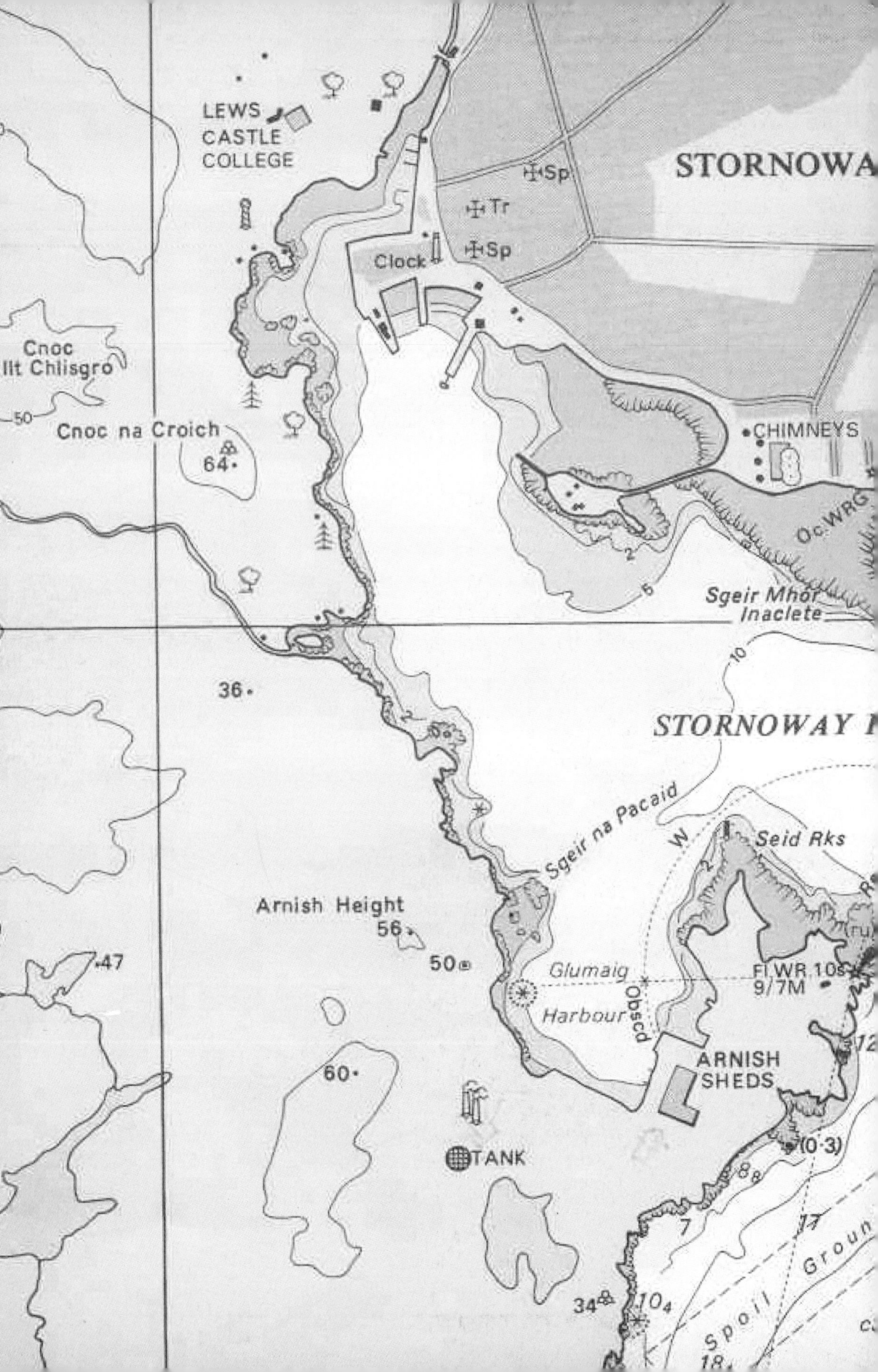
LEWS
CASTLE
COLLEGE
STORNOWA
Sp
Tr
Sp
Clock
Cnoc
Allt Chilisgro
50
Cnoc na Croich
64
CHIMNEYS
Oc WRG
2
5
Sgeir Mhor
Inaclete
10
36
STORNOWAY
Sgeir na Pacaid
W
Seid Rks
Arnish Height
56
50
47
Glumaig
Harbour
Obscd
Fl WR 10s
9/7M
ARNISH
SHEDS
60
TANK
(0·3)
7
17
Spoil
Groun
34
18